Technology Mining

Henrique da Hora · Alan L. Porter ·
Denise Chiavetta · Yi Zhang
Editors

Technology Mining

Text Analytics for Evidence-Based Foresight

Springer

Editors
Henrique da Hora
Instituto Federal Fluminense
Campos dos Goytacazes, Rio de Janeiro, Brazil

Denise Chiavetta
Search Technology, Inc.
Norcross, GA, USA

Alan L. Porter
Search Technology, Inc.
Norcross, GA, USA

Technology Policy and Assessment Center
Georgia Institute of Technology
Atlanta, GA, USA

Yi Zhang
Faculty of Engineering and Information Technology, Australian Artificial Intelligence Institute
University of Technology Sydney
Sydney, NSW, Australia

ISBN 978-3-032-10848-7 ISBN 978-3-032-10849-4 (eBook)
https://doi.org/10.1007/978-3-032-10849-4

This Springer imprint is published by the registered company Springer Nature Switzerland AG
The registered company address is: Gewerbestrasse 11, 6330 Cham, Switzerland

Preface

Tech Mining, emphasizing the practice of text analytics to derive empirical insights from Science, Technology, and Innovation (ST&I) information resources, has been a well-recognized methodological framework for decision support in ST&I policy and management (Porter & Cunningham, 2004). In the past two decades, we have observed many valuable Tech Mining studies, including:

- Extensive analyses of ST&I data from traditional bibliometric sources (e.g., research articles and patents) to various informetric datasets (e.g., news, social media, and policy documents) (Huang et al., 2016),
- Advanced methods incorporating artificial intelligence (AI) and data science techniques, such as intelligent bibliometrics (Zhang et al., 2020), and
- Diverse applications highlighting technology dynamics, interdisciplinary interactions, and broad societal, political, economic, and environmental impacts of emerging technologies (Carley et al., 2018; Huang et al., 2021).

As an annual event of the Tech Mining community, it has been more than 14 years since we organized the first Global Tech Mining (GTM) conference at Georgia Tech, Atlanta, the USA, in 2011. GTM has been physically held in the USA, Spain, and the Netherlands, and during the global pandemic, GTM conferences were run online with the local communities in China, Brazil, and Germany. We have observed a growing GTM community across different continents, countries, and regions—and across multiple disciplines, with prominent impacts to broad research fields.

With the major collection from the submissions of the GTM conferences, we have successfully published several special issues in prestigious journals in technology management and bibliometrics, such as *Technological Forecasting and Social Change* (Zhang et al., 2022), *IEEE Transactions on Engineering Management* (Zhang et al., 2021), *Scientometrics* (Zhang et al., 2017), and *Technology Analysis and Strategic Management* (Chiavetta & Porter, 2013). Selected papers were also archived in a book, *Anticipating Future Innovation Pathways Through Large Data Analysis* (Daim et al., 2016). A full list of all GTM special issues and book collections can be traced on the website: https://www.gtmconference.org/journal-special-issues/.

This book collects selected papers from the 12th GTM Conference (GTM2022) and some external submissions. Notably, GTM2022 is the second online GTM during the pandemic. It was jointly organized by the VP Institute and the GTM community in Brazil, led by the Oswaldo Cruz Foundation, the Federal Institute of Education, Science, Technology of Rio de Janeiro, and the Federal Institute of Education, Science, and Technology of Fluminense.

The ten chapters archived in this book were contributed by authors from seven countries, including Brazil, Japan, Iran, Mexico, Switzerland, the USA, and Finland. Their topics widely cross novel methodological developments of Tech Mining with modern artificial intelligence (AI) techniques, and empirical Tech Mining applications with real-world scenarios.

We were impressed by the interactive incorporation between Tech Mining and modern AI techniques. For example, Cassidy (2014) developed an active learning-based document classifier to support Tech Mining analysis. Valdecy et al. (2024) proposed a Python library for text summarisation, which integrates popular algorithms (e.g., TextRank) and some cutting-edge deep learning models (e.g., BART and ChatGPT). Tavazzi et al. (2024) proposed a set of proximity indices for measuring technological convergence and utilised a set of machine learning models for technological forecasting on encryption technologies, with OpenAlex.

Detecting emerging technologies has long been an interest of the GTM community. Iwami (2024) practiced Tech Mining approaches, including keyword scoring and clustering analysis, to identify emerging technologies in open-source software. Ashouri et al. (2024) proposed a novel large language model (LLM)-based method for detecting weak signals in quantitative foresight and examined the method in analyzing around 90,000 research articles. Vega-Gonzalez et al. (2024) utilised bibliometrics and technology roadmaps to review the research literature of surgery training, well practicing a hybrid Tech Mining solution with combined quantitative and qualitative approaches.

Extending Tech Mining's scope from newly emerging technologies to broad practical scenarios is always exciting, demonstrating Tech Mining's incredible potential in mining intelligence from rich ST&I data sources. Rabiei (2024) applied topic models to analyse more than 25,000 thesis records and over 440,000 search queries about energy research supply and demand in Iran's scientific repository. Darold and Ebersberger (2024) investigated the website data of over 79,000 German firms and revealed the correlations between the adoption of big data analytics and different firm features (including size, age, stakeholder, innovation activities, etc.). Carvalho (2024) implemented tech mining techniques, particularly text mining and ontologies, to criminal investigations, which targeted keyword-based behavioural and linguistic patterns in messages and electronic communications. Nakaya et al. (2024) investigated tolerance for uncertainty by analysing the transcripts of three meetings in seed-stage high-tech startups, in which text mining techniques and principal component analysis were involved.

Despite presenting a selective collection of most recent studies on Tech Mining, we have observed the increasing engagement of the community with AI, particularly modern machine learning techniques and large language models. As discussed in a revisit to Tech Mining, we anticipate a changing paradigm of Tech Mining due to AI's capabilities in information extraction, summarisation, and annotation (Porter et al., 2024). The involvement of LLMs in two chapters of this book would further endorse such a trend with AI-empowered Tech Mining. However, we still emphasise evidence-based foresight studies, with a convincing synergy of quantitative and qualitative approaches, and, thus, raise new questions about how we can responsibly practice AI models and AI-empowered indicators in technological forecasting and assessment, and conduct a robust, reliable, and explainable Tech Mining study (Zhang, 2023)?

Campos dos Goytacazes, Brazil — Henrique da Hora
Norcross/Atlanta, USA — Alan L. Porter
Norcross, USA — Denise Chiavetta
Sydney, Australia — Yi Zhang

References

Carley, S. F., Newman, N. C., Porter, A. L., & Garner, J. G. (2018). An indicator of technical emergence. *Scientometrics, 115*(1), 35–49.

Daim, T. U., Chiavetta, D., Porter, A. L., & Saritas, O. (2016). *Anticipating future innovation pathways through large data analysis*. Springer.

Huang, L., Chen, X., Ni, X., Liu, J., Cao, X., & Wang, C. (2021). Tracking the dynamics of co-word networks for emerging topic identification. *Technological Forecasting and Social Change, 170*, 120944.

Huang, Y., Youtie, J., Porter, A. L., Robinson, D. K., Cunningham, S. W., & Zhu, D. (2016). Big Data and Business: Tech mining to capture business interests and activities around Big Data. 2016 IEEE International Conferences on Big Data and Cloud Computing (BDCloud), Social Computing and Networking (SocialCom), Sustainable Computing and Communications (SustainCom) (BDCloud-SocialCom-SustainCom), Atlanta, GA, USA, 2016, pp. 145–150, https://doi.org/10.1109/BDCloud-SocialCom-SustainCom.2016.32

Porter, A. L., & Cunningham, S. W. (2004). *Tech mining: Exploiting new technologies for competitive advantage* (Vol. 29). John Wiley & Sons.

Porter, A. L., Zhang, Y., & Newman, N. C. (2024). Tech mining: a revisit and navigation. *Frontiers in Research Metrics and Analytics, 9*, 1364053.

Zhang, Y. (2023). Responsible models and indicators: challenges from artificial intelligence. In (Vol. 8, pp. 1305692): Frontiers Media SA.

Zhang, Y., Huang, Y., Chiavetta, D., & Porter, A. L. (2021). Guest Editorial: Tech mining for engineering management: An introduction. *IEEE Transactions on Engineering Management, 68*(5), 1211–1213.

Zhang, Y., Huang, Y., Chiavetta, D., & Porter, A. L. (2022). An introduction of advanced tech mining: Technical emergence indicators and measurements. *Technol. Forecast. Soc. Change, 182*(121855), 10.1016.

Zhang, Y., Porter, A., Cunningham, S. W., Chiavetta, D., & Newman, N. (2020). Parallel or intersecting lines? Intelligent bibliometrics for investigating the involvement of data science in policy analysis. *IEEE Transactions on Engineering Management, 68*(5), 1259–1271.

Zhang, Y., Porter, A. L., & Chiavetta, D. (2017). Scientometrics for tech mining: an introduction. *Scientometrics, 111*, 1875–1878.

Contents

Enhancing Tech Mining Analysis with Active Learning for Document Classification

Caitlin Cassidy

1 Introduction

This paper addresses an intersection between machine learning for text classification and the field of tech mining.

The automatic categorization of unstructured text is a subject of great interest to the machine learning community. As hardware capabilities become more advanced, so too do the systems designed to tackle such a task, and both are matched by a need for ever more data from which to learn. While the state of the art in text classification has inched closer to gold standard performance for decades, the amount of training required limits what is achievable.

Massive language models built on publicly available corpora are extremely useful for many text processing efforts, but they lack the subject matter expertise required to understand the nuance of any specific domain more narrow than, say, news text or general academic literature. Retaining the input of a human specialist, however, is expensive and time consuming, and in many cases, large amounts of training data introduce energy inefficiency and an increased carbon footprint. These constraints have inspired a new wave of models designed specifically to take on machine learning tasks using exceedingly small training corpora.

Such limited data necessitates an emphasis on the highest quality of human annotations, a process severely understudied within the discipline of machine learning. As this work's contribution, I propose an active learning framework designed to minimize the likelihood of annotator error and encourage ample coverage of even under-represented categories. In a case study mimicking the type of classification task that would require expert annotation, I show that this workflow is superior to random document selection in both regards.

C. Cassidy (✉)
Search Technology Inc., Norcross, Georgia
e-mail: caitlin.cassidy@searchtech.com

H. da Hora et al. (eds.), *Technology Mining*,
https://doi.org/10.1007/978-3-032-10849-4_1

Tech mining (Porter & Cunningham, 2004; Porter et al., to appear) combines bibliometric and text analyses to generate useful technical intelligence from Science, Technology and Innovation (ST&I) data resources, often addressing one or more sets of abstract records on a target R&D topic. For instance, in the study of "nanotechnology" research patterns (Rotolo et al., 2022), an essential first step is to distinguish what research is applicable to the topic (Wang et al., 2019b).

Though my focus here is the subsequent categorization of the resulting corpus of *records*, rather than the clustering of n-gram *terms* representing topics, the historical relevance of the latter necessitates a brief aside. Whereas term clustering encompasses ways to group topics for further analyses—e.g., latent semantic analysis (Landauer & Dumais, 1997) and topic modeling (Blei, 2012)—record clustering seeks to identify which records within the dataset "go together." To visualize, the software Vantage Point[1] offers a "cluster records" function that represents concentrations of like records as "haystacks from space."

When successful, such categorization contributes to analyzing research concentrations. To achieve those tech mining intents, I seek to facilitate the document-level classification process. Such a procedure follows one of three paradigms:

1. assigning records to existing, widely-used categories such as International Parent Classification (IPC) codes, which may too general for a given analysis;
2. grouping records using clustering techniques such as K-means (Duda et al., 1973) or DBSCAN (Ester et al., 1996), before assigning a label to each group (e.g., Yau et al. (2014)), which may result in clusters for which there is no clear label;
3. generating a novel categorization scheme to suit the topic at hand, which is the concern of the present article.

In most cases, such categorization starts with a human analyst manually assigning categories for each record. For a small record set, all records can so be labeled. For a large set (e.g., thousands—in the nanotechnology case, over a million), there is huge gain in having the machine learn assignment principles from the analyst's assigning a small set of records, with the computer completing classification of the remaining records, subject to analyst verification.

An essential step to such supervised automatic categorization is selecting the set of sample records for the analyst to assign, which follows one of two paths: "passive learning" by training the machine classifier on arbitrarily selected documents, or "active learning" by optimizing the selection of records for the human to assign. The selected documents are then used to train the machine learner to classify the remaining records.

This paper concerns the latter, an active learning paradigm which, had it been available, would have enabled analyses in the above-mentioned nanotechnology study that could not be done using ready-made categorization (such as the pre-assigned Web of Science Categories based on the journal or conference in which the research was published). As an example, one might study nanotechnology research concentrations

[1] https://thevantagepoint.com/

in relation to potential commercialization targets, seeking to identify cross-specialty knowledge transfer opportunities.

2 Background

Text Classification

Text classification is one of the oldest and most persistent tasks in machine learning. Early classifiers relied on carefully crafted rules based on expert linguistic knowledge. While such models were highly interpretable, they were extremely time-consuming to generate and had little to no usefulness outside of the specific tasks they were built to address.

This constraint is partially addressed by later systems which sought to generate and rank such rules automatically. Covering algorithms (Apte´ et al., 1994; Sasaki & Kita, 1998), decision trees (Chakrabarti et al., 2003; Johnson et al., 2002; Li & Jain, 1998), and association-based methods (Buddeewong & Kreesuradej, 2005; He & Loh, 2010) identify patterns between expert-indicated features and learn how best to apply them to new text. These approaches maintain the interpretability of their predecessors, but are generally outperformed by the subsequent wave of classification algorithms.

Supervised learners such as Naïve Bayes (Chen et al., 2009; Kim et al., 2006; Schneider, 2005), K-nearest Neighbors (Han et al., 2001; Kwon & Lee, 2003), and Support Vector Machines (Joachims, 1998; Mitra et al., 2007; Wang et al., 2006) take manually tagged training data as input, and learn characteristics of the categories from these examples. While these models eliminate the need for overt linguistic knowledge, they still require feature selection during pre-processing. And once again, compared to later generations of learners, their classification performance can be lackluster.

The state of the art in automatic text categorization employs Artificial Neural Networks (NNs), complex systems which pass vectorized inputs through multiple layers of nodes and edges, applying and adjusting aggregation functions at each layer. Inspired by biological neurons and synapses, these networks trace their origins to the Perceptron Learner (Gallant et al., 1990), and engineers have since taken advantage of advances in hardware to build "deep learners" such as Recurrent NNs (Arevian, 2007; Wang, 2018), Convolutional NNs (Lai et al., 2015; Wang et al., 2019a), Long Short-Term Memory models (Liu & Guo, 2019; Zhou et al., 2015, 2016), and many more, including hybrid approaches.

Deep learning models have the advantage of minimal human interference, having built-in mechanisms to convert raw text into appropriate input and learn optimal hyper-parameters, which makes such models fairly domain-agnostic and allows them flexibility in natural language and linguistic style. That said, they require an immense amount of training data for optimal performance, which is not readily available for small-scale and one-off projects.

Few-Shot Learning

The scarcity of training data for such applications has given rise to a class of machine learners dedicated to "few-shot" classification. Such models are designed to produce high-quality classifications with minimal training data—as few as five examples per category—and generally fall within one of three dominant paradigms.

One such family of learners (Sun et al., 2019; Ohashi et al., 2021) constructs a prototype representation of each category and compares it to new examples to determine how well they fit this abstraction of the class. Such an approach benefits from being self-contained and requiring no outside knowledge, making it domain-neutral and easily adapted for novel classification tasks. However, sparsity is still an issue.

Another archetype (Kim et al., 2022; Lee et al., 2021; Mukherjee & Awadallah, 2020) seeks to artificially inflate the amount of training data available, either through semi-supervised learning from unannotated documents with pseudo-labels, or by generating facsimiles with desirable features. Models employing these strategies are able to mitigate many of the obstacles posed by a lack of training examples, but may be prone to over-fitting to the imitation data.

Finally, many few-shot machine learners (Pan et al., 2019; Wang et al., 2021) employ a bootstrapping technique, whereby they refine models pre-trained on traditionally large corpora with data specific to the low-resource task. This method has the advantage of minimal parameterization, but can suffer from noise inadvertently introduced by the richer knowledge from outside the relevant domain.

Annotation Quality

With such limited training data to learn from, few-shot machine learners rely on high-quality annotations to perform competitively, a standard most commonly measured by the degree to which human raters agree on the labels assigned to a document. And indeed, a positive correlation between annotator agreement and classification performance is well-documented (Bermingham & Smeaton, 2009; Bobicev & Sokolova, 2017; Boguslav & Cohen, 2017; Brants, 2000; Teruel et al., 2018).

Though a low accuracy and agreement score may simply demonstrate the complexity inherent in a given categorization task, researchers often undertake efforts to improve agreement, nearly always by modifying annotation guidelines for clarity before re-tagging the training data. However, I argue that most small-scale projects of the kind requiring few-shot learners, specifically, are unlikely to employ multiple domain experts, or to have them re-visit annotation after updating the instructions.

Instead, some researchers have endeavored to curate the training data itself by guiding the annotation process. *Active Learning* is the process by which a machine learner selects unlabeled instances to be sent as queries to the annotator, often referred to in the literature as the *oracle*.

Settles (2009) divides existing active learning models into three main paradigms. In "membership query synthesis" (Alabdulmohsin et al., 2015; Schumann & Rehbein, 2019), the search space from which to select samples includes machine-generated pseudo-examples, artificially inflating the information available to the learner in much the same way as models discussed in Sect. 2.2. "Stream-based selective sampling" (Cacciarelli et al., 2022; Loy et al., 2012) allows the learner to decide online whether to query or discard each subsequent sample. It is important to note that this decision is agnostic to any other samples within the search space. "Pool-based active learning" (Joshi et al., 2009; Katz-Samuels et al., 2021; Nigam & McCallum, 1998; Zhan et al., 2021) is perhaps the most common method, due to the ease of collecting large, unlabeled datasets for training. Unlike stream-based selection, pool-based learning compares each candidate sample to others within the search space to select the most appropriate query.

In what remains of this manuscript, I propose a pool-based active learning approach for text classification focused on meeting two primary objectives: representing the full breadth of the search space by selecting diverse examples, and forming clear-cut categories by producing dense clusters of sample documents.

3 Methodology

The proposed workflow relies on computing similarities between examples in the training data and centroids representing each category and the corpus as a whole. Mechanically, documents are represented as Term Frequency-Inverse Document Frequency (TF-IDF, Salton and Buckley (1988)) vectors of each type (unique word) present in the example text. For any type w in a document d, the TF-IDF is given by Eqs. 1–3:

$$TF = \frac{Count(w, d)}{Length(d)} \tag{1}$$

$$IDF = \frac{Docs(total)}{Docs(relevant)} \tag{2}$$

$$TF{-}IDF = TF \times \log(IDF) \tag{3}$$

in which $Docs(total)$ indicates the total number of documents in the corpus (training and testing), and $Docs(relevant)$ indicates the number of documents in the corpus which contain at least one instance of the type. Prior to training, tokens (words in sequence) undergo Porter stemming (Porter, 1980), and common English stop words are removed.

As training progresses, centroids for each category, as well as the entire training set (the "total" centroid), are updated to reflect the "average" example of each. Each of these centroids is calculated to be the vector of the arithmetic mean of each type's TF-IDF across all examples previously assigned to the given category, or, in the case of the "total" centroid, all examples annotated thus far. These centroids can be said to represent what the machine learner knows about the data at a given point in training.

Guided Training

The active learning approach discussed here, which I refer to as *guided training*, begins by selecting at random a document with a number of tokens between the first and third quartiles of document length for the corpus. As each example is annotated, this window grows to allow both longer and shorter possible documents. This restriction ensures that training does not favor abnormally long or short examples.

Once the first document has been annotated, the second is the document least similar to the first example. Because the guided trainer's goal is to learn as much about the data as quickly as possible, it will prioritize documents that lie outside of its known search space, i.e., the least similar to those previously seen and trained. For each document classified, the total centroid and that of the assigned group are updated to reflect the new information.

Training continues by selecting the document least similar to the total centroid, constructed as an average vector across previous examples. This selection continues until all categories have been filled with at least three example documents. At this point, the guide switches to presenting the annotator with the document most similar to the least populated group. Once the learner has an understanding of where each category lies in the search space, it seeks to flesh out its knowledge by expanding outward from the center of each class.

This process continues until the desired number of training examples is reached, as illustrated in Fig. 1. As training progresses, the learner may gain enough confidence to recommend a category for the current document. The assigned annotation then informs the learner if its intuition was correct. By tracking the precision and recall of these machine guesses compared to their official classification, the interface can also indicate the confidence level of the system as a whole, rather than just the likelihood that a given example belongs to a given category.

(a) Begin by selecting a random document.

(b) The next document is the one least similar to the initial example.

(c) Then the least similar to the previous two, then previous three, and so on.

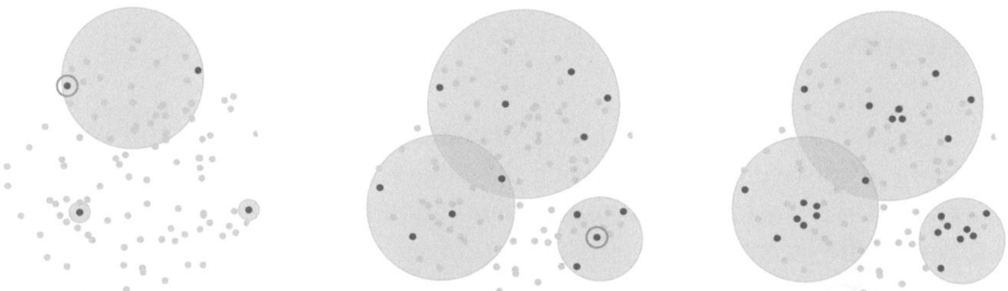

(d) Documents in the same category share a centroid, which is the averaged vector of all constituents.

(e) Once we have three in each category, select most similar to centroid of least populated category.

(f) Continue until desired *n* documents.

Fig. 1 Guided training in two-dimensional space with three categories

Data

To demonstrate the efficacy of guided training over a random shuffle, two corpora were constructed from biomedical data with closely tied and easily separable categories, respectively. Both were limited to a publication year between 2010 and 2020, and non-English documents and those with an empty title or abstract were removed. The closely tied corpus, which I will call the Narrow Corpus, contains 20,128 titles and abstracts from a PubMed search of five MeSH categories in dentistry, 536 of which belong to more than one category. It has a vocabulary density of and a median document length of 145 tokens. The corpus with more distinct categories, the Broad Corpus, contains 24,855 documents from a search in five unrelated categories, with a density of 53.62 and median length of 137 tokens. Twelve documents were retrieved with more than one search. The MeSH categories and distributions for both corpora are presented in Table 1.

Table 1 Distribution of documents in demonstration corpora

(a) Narrow corpus		(b) Broad corpus	
Class	Documents	Class	Documents
Orthodontics	10,520	Cyanobacteria	9,338
Endodontics	5,096	Cataract	6,197
Periodontics	2,353	Root canal	5,096
Public health	1,558	Biocatalysis	2,206
Oral surgery	1,173	Veterinary medicine	2,030

4 Results

The algorithm described in Sect. 3 is included in Vantage Point software, and the following results were obtained using the Smart-Trainer function. For each corpus, 100 records were selected for human annotation using both guided training and a random shuffle (active versus passive learning).

I use several metrics as a proxy for measuring the likelihood of annotator error. One is the difference between the lowest-similarity category to which the example is assigned and the highest-similarity category to which it is not, averaged across the training run. A higher value in this metric indicates that, on average, the documents presented for training are distinctly more similar to their assigned classes than to the others. These values for both corpora and training styles are presented in Table 2.

For both corpora, guided training widened the gap in similarity between positive and negative categories by a large margin, more than four times (0.129 versus 0.028) within the Broad Corpus. This indicates that guided training is leading annotators to categorize more clear-cut examples than its random counterpart, which should ultimately reduce error.

Of note, however, is that the median difference is lower in the Narrow Corpus. This, coupled with the much higher standard deviations, shows that guided training is more prone to variation in distinction, with very pronounced differences skewing the mean higher. Regardless, the presence of these highly distinct examples demonstrates the efficacy of guided training for annotating unambiguous documents.

A similar metric is the difference between in- and out-of-class examples by category. For a given category, I take the mean similarity (at time of annotation) for all documents assigned to said class, and compare to the similarity for documents

Table 2 Difference between lowest in-class and highest out-of-class similarities

(a) Narrow corpus			(b) Broad corpus		
	Guided	Random		Guided	Random
Mean	0.056	0.023	Mean	0.129	0.028
Median	0.005	0.020	Median	0.058	0.022
St. Dev	0.140	0.034	St. Dev	0.017	0.034

which were excluded. The difference in these values indicates how easily separable documents assigned to a category are from other classes. Results are displayed in Table 3, with highest average in-class and lowest average out-of-class similarities highlighted in bold.

For both corpora, mean in-class similarity is higher across the board with guided training, as is the difference between in- and out-of-class similarities. In fact, the difference using guided training is nearly three times that of random training in the Narrow Corpus, and almost four times higher in the Broad Corpus. This shows once again that guided training is prioritizing documents with very clear-cut category membership. Note, however, that this difference can vary markedly between classes. In the Narrow Corpus, for example, Public Health is quite easy to differentiate from other classes, having the second-highest similarity difference using both guided and random training. However, the boundaries of Periodontics are made nearly ten times more explicit with guided training, proving this method particularly useful for training underrepresented categories.

We can compute the difference in median in- and out-of-class similarity as well. Table 4 shows much the same pattern, with the exception that the Narrow Corpus favors random training for Orthodontics and Endodontics. Though it is surprising to see a result favoring passive learning, it represents yet more indication that guided training is most effective for classes that make up a smaller proportion of the data, while demonstrating the bias toward very high-similarity documents. Such documents skew the mean difference higher than the median when using guided training. The law of large numbers may also be at play: because Orthodontics is so dominantly

Table 3 Mean centroid similarity to in-class versus out-of-class training documents by category

(a) Narrow corpus, Guided				(b) Narrow corpus, random			
Class	In	Out	Diff.	Class	In	Out	Diff.
Orthodontics	0.072	0.028	0.044	Orthodontics	0.061	0.034	0.027
Endodontics	0.107	**0.014**	0.094	Endodontics	**0.080**	0.030	**0.050**
Periodontics	**0.234**	0.025	**0.209**	Periodontics	0.046	0.025	0.021
Oral surgery	0.107	0.038	0.068	Oral surgery	0.021	**0.011**	0.010
Public health	0.204	0.046	0.158	Public health	0.075	0.028	0.046
Overall	0.144	0.030	0.114	Overall	0.063	0.024	0.039
(c) Broad corpus, guided				**(d) Broad corpus, random**			
Class	In	Out	Diff.	Class	In	Out	Diff.
Cyanobacteria	0.051	0.011	0.039	Cyanobacteria	0.044	0.017	0.028
Cataract	0.292	**0.007**	0.285	Cataract	0.055	0.017	0.038
Biocatalysis	0.068	0.036	0.031	Biocatalysis	0.018	**0.007**	0.010
Root canal	**0.303**	0.009	**0.294**	Root Canal	**0.084**	0.019	**0.065**
Vet. Med	0.199	0.008	0.192	Vet. Med	0.044	0.012	0.032
Overall	0.168	0.014	0.154	Overall	0.053	0.014	0.039

represented in the dataset, guided training may be ignoring a large segment of the class by homing in on the records closest to the first three examples discovered during initialization. Essentially, the larger categories may be falling prey to over-fitting.

Additionally, I show that guided training populates each category more quickly than random selection. In order to learn effectively, the model must have at least a few examples in each class. However, human annotators are costly and prone to fatigue, so it is beneficial for the learner to build an understanding of all classes as quickly (i.e., in as few examples) as possible. Table 5 shows the number of manual classifications needed to populate each category with n examples, with the maximum representing the point at which *all* classes achieve that number.

For both corpora, guided training populates the categories in fewer examples than random selection. As in all other regards, the benefits of guided training are most pronounced for underrepresented categories. Indeed, in the Narrow Corpus, a random training order never produces more than three documents for least-common Oral Surgery to learn from. The result of which is an underdeveloped picture of this class by the time the annotator reaches fatigue. To ensure adequate training material within the constraints of human attention, guided training is yet again preferable.

Finally, I compare the learning curves of these active learning models for both corpora. Figure 2 displays precision, recall, and F1 measure for automatically categorizing a withheld test set using both guided and random document selection for training sets ranging from 20 to 50 documents. The test set is comprised of all documents not selected by either model, 20,028 records in the narrow corpus and 24,754 in the broad. The classifier used is a cosine similarity-based model available in Vantage Point software.

Table 4 Median centroid similarity to in-class versus out-of-class training documents by category

(a) Narrow corpus, guided				(b) Narrow corpus, random			
Class	In	Out	Diff	Class	In	Out	Diff
Orthodontics	0.037	0.018	0.019	Orthodontics	0.064	0.031	0.033
Endodontics	0.020	0.008	0.011	Endodontics	**0.089**	0.028	**0.060**
Periodontics	0.255	**0.007**	**0.248**	Periodontics	0.049	0.023	0.026
Oral surgery	0.115	0.012	0.103	Oral surgery	0.024	**0.009**	0.015
Public health	**0.260**	0.014	0.246	Public health	0.059	0.022	0.038
Overall	0.115	0.011	0.105	Overall	0.062	0.021	0.041
(c) Broad corpus, guided				(d) Broad corpus, random			
Class	In	Out	Diff	Class	In	Out	Diff
Cyanobacteria	0.059	0.008	0.051	Cyanobacteria	0.042	0.016	0.026
Cataract	0.335	0.006	0.329	Cataract	0.052	0.014	0.038
Biocatalysis	0.085	0.007	0.077	Biocatalysis	0.020	**0.005**	0.015
Root Canal	**0.369**	0.008	**0.361**	Root Canal	**0.070**	0.015	**0.055**
Vet. Med	0.176	**0.004**	0.172	Vet. Med	0.035	0.010	0.025
Overall	0.087	0.007	0.081	Overall	0.045	0.011	0.034

Table 5 Number of annotations needed to classify *n* examples in each category

(a) Narrow corpus, guided

Class	1	2	3	4	5
Orthodontics	3	5	10	12	17
Endodontics	2	7	8	9	11
Periodontics	15	16	27	28	30
Oral surgery	4	19	22	26	29
Public health	1	6	15	20	25
Max	15	19	27	28	30

(b) Narrow corpus, random

Class	1	2	3	4	5
Orthodontics	1	2	3	8	9
Endodontics	5	7	12	13	24
Periodontics	4	6	16	18	62
Oral surgery	23	56	89	-	-
Public health	8	15	31	33	37
Max	23	56	89	-	-

(c) Broad corpus, guided

Class	1	2	3	4	5
Cyanobacteria	4	5	9	11	16
Cataract	3	14	18	19	44
Biocatalysis	2	12	13	17	43
Root Canal	1	6	7	20	23
Vet. Med	8	10	15	21	22
Max	8	14	18	21	44

(d) Broad corpus, random

Class	1	2	3	4	5
Cyanobacteria	6	7	9	11	13
Cataract	1	2	10	12	14
Biocatalysis	18	19	33	81	87
Root Canal	3	4	5	15	31
Vet. Med	8	22	24	27	35
Max	18	22	33	81	87

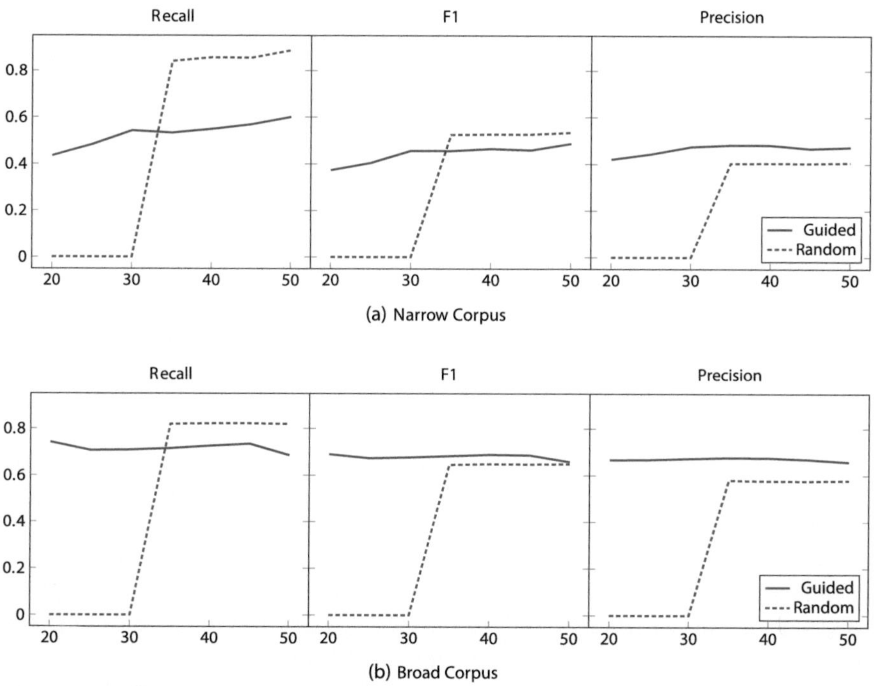

Fig. 2 Learning curve by training set size (20–50 documents)

Unsurprisingly, the broad corpus is easier to categorize across all metrics, but it is interesting to note that with guided training, the narrow corpus shows a steady increase in performance with a greater number of selected documents, while the broad corpus starts very strong and gradually decreases. This implies an over-fitting issue, which indicates that this active learning paradigm is more appropriate for corpora with a higher degree of overlap.

I also note the difference between the patterns for precision and recall. In both corpora, all three metrics begin higher for guided training, while random recall surpasses guided recall after 30 records, around the point at which guided training is considered "initialized" with three records per category. This is likely because the "fleshing out" stage of guided training homes in on each category, increasing the threshold for inclusion in any given class. Random training has no such focus and is thus less "picky". For the same reason, precision remains higher with guided training.

In any case, F1, in balancing the twin goals of finding all and only true positives, appears to converge between the active learning models. This supports the efficacy of guided training for producing a usable training corpus as quickly as possible. While random document selection may produce stronger results with more training records, in small-scale or one-off projects, categorization as one part of a larger analysis benefits from the rapid training supported by the guided learning model.

5 Conclusion and Future Work

While new models of machine learning are being churned out every day to compete for supremacy in benchmark text classification tasks, the process of creating these training corpora is clearly under-studied. Many such annotations undergo multiple rounds of guideline refinement, using time and resources in the process.

I suggest instead that a guided training workflow which presents examples to annotators in an intentional order decreases the likelihood of mis-classification and populates each category in fewer examples. This intuition is borne out in the results of a case study using biomedical text and trained on 100 documents across five categories, though it is not without caveats.

One is the degree to which the initial document controls the rest of the workflow. By randomly selecting a starting point, we run the risk of never encountering extremely useful examples, though the same could be said of a totally arbitrary training order. A refinement of the guided training process should address a heuristic for finding the best opening document. A very straightforward option is simply to begin with the median-length record, which doesn't necessarily mitigate the problem, as this record is as likely as any other to mislead the search.

An alternative would be to consider the search space as a whole, either by selecting the document most or least similar to the centroid of the entire record pool. This line of inquiry merits empirical study, as both provide theoretical benefits and pitfalls. Beginning with the most central document could result in a training corpus more representative of the typical data in a search space, but may struggle to meaningfully encode edge cases. Likewise, seeding training with the most peripheral document may cover more of the search space to the detriment of the more densely clustered examples closer to the centroid. This runs the risk of missing the valuable information provided by documents that sit at the boundary between categories. Experimentation of these heuristics, as well as cross-validation using different random seeds, would be an illuminating future study.

The primary advantage of this training paradigm can also be something of a detriment. After the initialization phase of finding more and more dissimilar documents, the focus shifts to fleshing out each category, resulting in a set of examples concentrated near the centers of each class's search space. This all but ignores edge cases and the useful information they likely contain. However, that information is often accompanied or overshadowed by noise, so there is still merit in keeping to more clear-cut data.

Another potential variable is the use of TF-IDF for abstracting text strings into numerical data. While other document embedding techniques are available (Kiros et al., 2015; Le & Mikolov, 2014), this model was designed to be as domain-agnostic as possible. Because pre-trained embeddings are generated using data that is likely not analogous to the corpus being studied, they are inappropriate for the niche, one-off projects targeted by this technique. Similarly, training these models for the specific project requires large amounts of time and data which may not be available. That said, for more commonly studied text domains, such as news articles or academic text, a comparison between TF-IDF and more elaborate embedding techniques is warranted.

The scope of this paper is also limited to the results of two case studies constructed specifically to mimic the experience of a one-time expert annotator. Further study is needed to expand the applicability to other domains. Likewise, a direct comparison to other active learning paradigms, such as those discussed in Sect. 2.3, is needed to demonstrate robustness.

While this guided training workflow is designed for a single annotator with a one-off classification scheme, further validation in the form of a rater agreement study is warranted. Because high agreement is correlated with better performance, demonstrating improved annotation uniformity over random document selection would further support the usefulness of this system.

Other aspects of the annotator experience, such as category recommendations and an indication of the learner's confidence, while not explored here, may be able to reduce fatigue or imply a critical mass in number of examples. This would allow researchers to limit the time required of a subject matter expert while also building a robust training corpus.

Acknowledgements Special thanks to Alan L. Porter and Nils C. Newman of Search Technology, Inc. for sharing their Tech Mining expertise, as well as data collection and indispensable feedback during the writing process.

References

Alabdulmohsin, I., Gao, X., & Zhang, X. (2015). Efficient active learning of halfspaces via query synthesis. In *Proceedings of the AAAI Conference on Artificial Intelligence* (Vol. 29).
Apté, C., Damerau, F., & Weiss, S. M. (1994). Automated learning of decision rules for text categorization. *ACM Transactions on Information Systems (TOIS), 12*(3), 233–251.
Arevian, G. (2007). Recurrent neural networks for robust real-world text classification. In *IEEE/WIC/ACM International Conference on Web Intelligence (WI'07), IEEE*, (pp. 326–329).
Bermingham, A., & Smeaton, A. F. (2009). A study of inter-annotator agreement for opinion retrieval. In *Proceedings of the 32nd International ACM SIGIR Conference on Research and Development in Information Retrieval* (pp. 784–785).
Blei, D. M. (2012). Probabilistic topic models. *Communications of the ACM, 55*(4), 77–84.
Bobicev, V., & Sokolova, M. (2017). Inter-annotator agreement in sentiment analysis: Machine learning perspective. In *International Conference Recent Advances in Natural Language Processing*, (pp. 97–102).
Boguslav, M., & Cohen, K. B. (2017). Inter-annotator agreement and the upper limit on machine performance: Evidence from biomedical natural language processing. *Studies in Health Technology and Informatics, 245*, 298–302.
Brants, T. (2000). Inter-annotator agreement for a German newspaper corpus. In LREC, Citeseer.
Buddeewong, S., & Kreesuradej, W. (2005). A new association rule-based text classifier algorithm. In *17th IEEE International Conference on Tools with Artificial Intelligence (ICTAI'05), IEEE* (p. 2).
Cacciarelli, D., Kulahci, M., & Tyssedal, J. S. (2022). Stream-based active learning with linear models. *Knowledge-Based Systems, 254*, Article 109664.
Chakrabarti, S., Roy, S., & Soundalgekar, M. V. (2003). Fast and accurate text classification via multiple linear discriminant projections. *The VLDB Journal, 12*, 170–185.
Chen, J., Huang, H., Tian, S., & Qu, Y. (2009). Feature selection for text classification with Naïve Bayes. *Expert Systems with Applications, 36*(3), 5432–5435.
Duda, R. O., & Hart, P. E. (1973). *Pattern classification and scene analysis* (Vol. 3). Wiley.
Ester, M., Kriegel, H. P., Sander, J., & Xu, X. (1996). A density-based algorithm for discovering clusters in large spatial databases with noise. *KDD, 96*, 226–231.
Gallant, S. I. (1990). Perceptron-based learning algorithms. *IEEE Transactions on Neural Networks, 1*(2), 179–191.
Han, E. H., Karypis, G., & Kumar, V. (2001). Text categorization using weight adjusted k-nearest neighbor classification. In *Advances in Knowledge Discovery and Data Mining: 5th Pacific-Asia Conference, PAKDD 2001 Hong Kong, China, April 16–18, 2001 Proceedings 5* (pp. 53–65). Springer.
He, C., & Loh, H. T. (2010). Pattern-oriented associative rule-based patent classification. *Expert Systems with Applications, 37*(3), 2395–2404.
Joachims, T. (1998). Text categorization with support vector machines: Learning with many relevant features. In *European Conference on Machine Learning* (pp. 137–142). Springer.
Johnson, D. E., Oles, F. J., Zhang, T., & Goetz, T. (2002). A decision-tree-based symbolic rule induction system for text categorization. *IBM Systems Journal, 41*(3), 428–437.
Joshi, A. J., Porikli, F., & Papanikolopoulos, N. (2009). Multi-class active learning for image classification. In *2009 IEEE conference on computer vision and pattern recognition*, IEEE (pp. 2372–2379).
Katz-Samuels, J., Zhang, J., Jain, L., & Jamieson, K. (2021). Improved algorithms for agnostic pool-based active classification. In *International Conference on Machine Learning, PMLR* (pp. 5334–5344).
Kim, H. H., Woo, D., Oh, S. J., Cha, J. W., & Han, Y. S. (2022). Alp: Data augmentation using lexicalized PCFGs for few-shot text classification. *Proceedings of the AAAI Conference on Artificial Intelligence, 36*, 10894–10902.

Kim, S. B., Han, K. S., Rim, H. C., & Myaeng, S. H. (2006). Some effective techniques for naive bayes text classification. *IEEE Transactions on Knowledge and Data Engineering, 18*(11), 1457–1466.

Kiros, R., Zhu, Y., Salakhutdinov, R. R., Zemel, R., Urtasun, R., Torralba, A., & Fidler, S. (2015). Skip-thought vectors. Advances in neural information processing systems 28.

Kwon, O. W., & Lee, J. H. (2003). Text categorization based on k-nearest neighbor approach for web site classification. *Information Processing & Management, 39*(1), 25–44.

Lai S, Xu L, Liu K, & Zhao J (2015) Recurrent convolutional neural networks for text classification. In *Proceedings of the AAAI Conference on Artificial Intelligence* (vol 29).

Landauer, T. K., & Dumais, S. T. (1997). A solution to Plato's problem: The latent semantic analysis theory of acquisition, induction, and representation of knowledge. *Psychological Review, 104*(2), 211.

Le, Q., & Mikolov, T. (2014). Distributed representations of sentences and documents. In *International Conference on Machine Learning*, PMLR (pp. 1188–1196).

Lee, J. H., Ko, S. K., & Han, Y. S. (2021). Salnet: Semi-supervised few-shot text classification with attention-based lexicon construction. *Proceedings of the AAAI Conference on Artificial Intelligence, 35*, 13189–13197.

Li, Y. H., & Jain, A. K. (1998). Classification of text documents. *The Computer Journal, 41*(8), 537–546.

Liu, G., & Guo, J. (2019). Bidirectional LSTM with attention mechanism and convolutional layer for text classification. *Neurocomputing, 337*, 325–338.

Loy, C. C., Hospedales, T. M., Xiang, T., & Gong, S. (2012). Stream-based joint exploration–exploitation active learning. In *2012 IEEE Conference on Computer Vision and Pattern Recognition, IEEE* (pp. 1560–1567).

Mitra, V., Wang, C. J., & Banerjee, S. (2007). Text classification: A least square support vector machine approach. *Applied Soft Computing, 7*(3), 908–914.

Mukherjee, S., & Awadallah, A. (2020). Uncertainty-aware self-training for few-shot text classification. *Advances in Neural Information Processing Systems, 33*, 21199–21212.

Nigam, K., & McCallum, A. (1998). Pool-based active learning for text classification. In *Conference on Automated Learning and Discovery (CONALD), Citeseer*.

Ohashi, S., Takayama, J., Kajiwara, T., & Arase, Y. (2021). Distinct label representations for few-shot text classification. In *Proceedings of the 59th Annual Meeting of the Association for Computational Linguistics and the 11th International Joint Conference on Natural Language Processing* (Vol. 2: Short Papers, pp. 831–836).

Pan, C., Huang, J., Gong, J., & Yuan, X. (2019). Few-shot transfer learning for text classification with lightweight word embedding based models. *IEEE Access, 7*, 53296–53304.

Porter, A., Zhang, Y., & Newman, N. (To appear) Tech mining: A revisit and navigation. Frontiers.

Porter, A. L., & Cunningham, S. W. (2004). *Tech mining: Exploiting new technologies for competitive advantage*. John Wiley & Sons.

Porter, M. F. (1980). An algorithm for suffix stripping. *Program, 14*(3), 130–137.

Rotolo, D., Natalicchio, A., Porter, A., & Schoeneck, D. J. (2022). The emergence of inter-organisational networks in emerging technologies: The case of microneedles. In *Global Tech Mining Conference*.

Salton, G., & Buckley, C. (1988). Term-weighting approaches in automatic text retrieval. *Information Processing & Management, 24*(5), 513–523.

Sasaki, M., & Kita, K. (1998). Rule-based text categorization using hierarchical categories. In *SMC'98 Conference Proceedings*. In *1998 IEEE International Conference on Systems, Man, and Cybernetics (Cat. No. 98CH36218), IEEE* (Vol. 3, pp. 2827–2830).

Schneider, K. M. (2005). Techniques for improving the performance of naive bayes for text classification. In *Computational Linguistics and Intelligent Text Processing: 6th International Conference, CICLing 2005, Mexico City, Mexico, February 13–19, 2005. Proceedings 6* (pp. 682–693). Springer.

Schumann, R., & Rehbein, I. (2019). Active learning via membership query synthesis for semi-supervised sentence classification. In *Proceedings of the 23rd Conference on Computational Natural Language Learning (CoNLL)* (pp. 472–481).
Settles, B. (2009). *Active learning literature survey*. Computer Sciences Technical Report 1648, University of Wisconsin–Madison.
Sun, S., Sun, Q., Zhou, K., & Lv, T. (2019). Hierarchical attention prototypical networks for few-shot text classification. In *Proceedings of the 2019 Conference on Empirical Methods in Natural Language Processing and the 9th International Joint Conference on Natural Language Processing* (EMNLP-IJCNLP) (pp. 476–485).
Teruel, M., Cardellino, C., Cardellino, F., Alemany, L. A., & Villata, S. (2018). Increasing argument annotation reproducibility by using inter-annotator agreement to improve guidelines. In *Proceedings of the Eleventh International Conference on Language Resources and Evaluation (LREC 2018)*.
Wang, B. (2018). Disconnected recurrent neural networks for text categorization. In *Proceedings of the 56th Annual Meeting of the Association for Computational Linguistics* (Vol. 1: Long Papers, pp. 2311–2320).
Wang, J., Wang, K. C., Rudzicz, F., & Brudno, M. (2021). Grad2task: Improved few-shot text classification using gradients for task representation. *Advances in Neural Information Processing Systems, 34*, 6542–6554.
Wang, R., Li, Z., Cao, J., Chen, T., & Wang, L. (2019a). Convolutional recurrent neural networks for text classification. In *2019 International Joint Conference on Neural Networks (IJCNN)*, IEEE (pp. 1–6).
Wang, Z., Porter, A. L., Kwon, S., Youtie, J., Shapira, P., Carley, S. F., & Liu, X. (2019b). Updating a search strategy to track emerging nanotechnologies. *Journal of Nanoparticle Research, 21*, 1–21.
Wang, Z. Q., Sun, X., Zhang, D. X., & Li, X. (2006). An optimal SVM-based text classification algorithm. In *2006 International Conference on Machine Learning and Cybernetics, IEEE* (pp. 1378–1381).
Yau, C. K., Porter, A., Newman, N., & Suominen, A. (2014). Clustering scientific documents with topic modeling. *Scientometrics, 100*, 767–786.
Zhan, X., Liu, H., Li, Q., & Chan, A. B. (2021). A comparative survey: Benchmarking for pool-based active learning. In *IJCAI* (pp. 4679–4686).
Zhou, C., Sun, C., Liu, Z., & Lau, F. (2015). A c-lstm neural network for text classification. arXiv:151108630.
Zhou, P., Qi, Z., Zheng, S., Xu, J., Bao, H., & Xu, B. (2016). Text classification improved by integrating bidirectional LSTM with two-dimensional max pooling. arXiv:161106639.

Unveiling pyAutoSummarizer: An Extractive and Abstractive Summarization Library Powered with Artificial Intelligence

Valdecy Pereira, Rodrigo Celso de Lima Porto, Luiz Antonio Aredes Figueira, and Rafael Alves Costa Araujo Ferreira

1 Introduction

Natural Language Processing (NLP) has experienced remarkable advancements due to the proliferation of textual data and the demand for sophisticated automated systems capable of comprehending and extracting meaning from it. NLP addresses various challenges, including sentiment analysis, named entity recognition, machine translation, question answering, and more (Hossain et al., 2023; Collobert et al., 2011). Nevertheless, one of the pivotal problems NLP has tackled is the intricate task of text summarization, which aims to distill large volumes of information into concise and coherent summaries, enabling users to grasp the main points and essence of the text efficiently.

Text summarization involves creating a concise and coherent summary of a longer text, maintaining the essential meaning and information contained within the original document. The need for text summarization has become increasingly evident with the influx of information in our era. Condensing lengthy texts into brief, informative summaries is crucial for efficient information retrieval, comprehension, and decision-making (Multu & Sezer, 2023). However, summarizing text is not a trivial task. It requires an understanding of the text's semantic structure, the ability to identify

V. Pereira (✉) · R. C. de Lima Porto · L. A. A. Figueira · R. A. C. A. Ferreira
Department of Production Engineering, Federal Fluminense University, Niterói, Brazil
e-mail: valdecypereira@id.uff.br

R. C. de Lima Porto
e-mail: rodrigo_celso@id.uff.br

L. A. A. Figueira
e-mail: luizfigueira@id.uff.br

R. A. C. A. Ferreira
e-mail: rafaelacaf@id.uff.br

H. da Hora et al. (eds.), *Technology Mining*,
https://doi.org/10.1007/978-3-032-10849-4_2

critical points, and the ability to succinctly express them without losing context or meaning.

According to Gambhir and Gupta (2017), two major approaches to text summarization exist: extractive and abstractive. Extractive summarization involves identifying and extracting key phrases or sentences from the original text to construct a summary. In contrast, abstractive summarization goes a step further; it consists of understanding the text and generating a new summary, sometimes using wording not present in the original text. This approach is more complex as it requires a higher level of understanding and language generation capability.

To simplify the complex text summarization task in NLP, we present pyAutoSummarizer, a comprehensive Python library that effortlessly generates concise and meaningful summaries from textual data. pyAutoSummarizer is designed to address the complexities of text summarization; with a comprehensive approach, this library combines extractive and abstractive summarization techniques, incorporating classical methods and state-of-the-art deep learning algorithms. For extractive summarization, it includes TextRank (Mihalcea & Tarau, 2004), LexRank (Erkan & Radev, 2004), LSA—Latent Semantic Analysis (Steinberger & Karel, 2004), KL-Sum—Kullback-Leibler Sum (Haghighi & Vanderwende, 2009), and the deep-learning models BART—Bidirectional and Auto-Regressive Transformers (Lewis et al., 2019) and T5—Text-to-Text Transfer Transformer (Raffeel et al., 2020). T5, known for its abstractive summarization capabilities, can sometimes tend to directly copy exact sentences from the original text, especially in shorter texts. This unique characteristic makes it suitable for integration into extractive algorithms, as it combines elements of both extractive and abstractive approaches.

Concerning abstractive summarization, pyAutoSummarizer introduces state-of-the-art deep-learning models that push the boundaries of text generation. Included in our comprehensive library are PEGASUS (Pre-training with Extracted Gap-sentences for Abstractive Summarization), a groundbreaking model developed by Zhang et al. in 2020a, 2020b and OpenAI's remarkable generative powerhouse, ChatGPT (Openai, 2023). These cutting-edge models unleash the potential for generating concise and coherent summaries that capture the essence of the input text, revolutionizing the field of abstractive summarization.

pyAutoSummarizer also has robust preprocessing features, recognizing the necessity of effective preprocessing for high-quality text analysis; the library is equipped with various functionalities that aid in preparing text data. Firstly, it supports stop words removal in a multitude of languages, including Arabic, Bengali, Bulgarian, Chinese, Czech, English, Finnish, French, German, Greek, Hebrew, Hind, Hungarian, Italian, Japanese, Korean, Marathi, Persia, Polish, Portuguese-br, Romanian, Russian, Slovak, Spanish, Swedish, Thai, and Ukrainian. This feature helps filter out the noise associated with frequently occurring words with little semantic value, allowing the algorithms to focus on the text's essential aspects (GAG & SHAH, 2015). In addition, the library can convert text to lowercase, remove special characters, remove numbers, and remove custom words that could interfere with the summarization process. This capability addresses the challenges in preparing text data for effective summarization.

To ensure the quality of its summaries, pyAutoSummarizer integrates a suite of evaluation metrics, comparing the generated summary against a reference. These include Rouge-N (Lin & Hovy, 2003), Rouge-L (Lin & OCH, 2004), and Rouge-S (Lin & Och, 2004), which look at the overlap of n-grams, longest common subsequence, and skip-bigram between the generated summary and reference, respectively. Other evaluation tools, such as BLEU—Bilingual Evaluation Understudy (Papineni et al., 2002) and METEOR—Metric for Evaluation of Translation with Explicit Ordering (Banerjee & Lavie, 2005), are also incorporated.

PyAutoSummarizer differentiates itself in the field of text summarization through a unique combination of extractive and abstractive methods, leveraging algorithms such as TextRank, LexRank, LSA, and KL-Sum, in addition to state-of-the-art machine learning models like BART, T5, PEGASUS, and ChatGPT. This integration enables it to produce summaries that are not only contextually relevant but also critically assessed through a suite of evaluation metrics, including Rouge-N, Rouge-L, Rouge-S, BLEU, and METEOR. Furthermore, PyAutoSummarizer enhances accessibility across diverse linguistic environments with text normalization and multi-language stopword support. In contrast, tools like Sumy (MISO, 2023) and Gensim (Řehurek & Sojka, 2011), while proficient in extractive summarization, lack the abstractive capabilities and the broad range of evaluation metrics that PyAutoSummarizer offers, limiting their utility in more complex summarization scenarios that benefit from paraphrasing and nuanced language generation. Specifically, Sumy's reliance on algorithms such as Luhn (LUHN, 1958) and Edmundson (Edmundson, 1969) restricts its applicability to extractive summarization. Gensim, despite its strengths in topic modeling and text similarity, falls short of providing the abstractive summarization, advanced models, and comprehensive evaluation metrics found in PyAutoSummarizer, making the latter a more versatile and effective solution for summarization tasks.

In Sect. 1, we introduce the problem of text summarization in NLP and position pyAutoSummarizer as a potential comprehensive solution. Section 2 explores the core functionalities of the library, including summarization techniques, evaluation metrics, and usability features. Section 3 showcases a case study highlighting the practical application and benefits of pyAutoSummarizer. Section 4 evaluates the strengths and limitations of pyAutoSummarizer, considering its range of techniques and integration of evaluation metrics. Finally, Sect. 5 summarizes the key findings and emphasizes the significance of pyAutoSummarizer in automating and improving text summarization.

2 The Core Functionalities of pyAutoSummarizer

pyAutoSummarizer has a robust repertoire of algorithms designed to tackle text summarization effectively. These algorithms can be broadly classified into Extractive and Abstractive Summarization Techniques. To initialize pyAutoSummarizer, begin

```
# Create Instance
smr = summarization(text_1, stop_words = ['en'] , n_words = -1, n_chars = -1, lowercase =
True, rmv_accent = True, rmv_special_chars = True, rmv_numbers = True,
rmv_custom_words = [ ], verbose = True )
```

Fig. 1 Instance

by creating an instance. For illustration, in Fig. 1, we'll construct an instance named "smr":

In this configuration, *text_1* represents the string containing the text to be summarized. The *stop_words* parameter is a list specifying languages for stopwords removal; "ar" = Arabic; "bn" = Bengali; "bg" = Bulgarian; "cs" = Czech; "en" = English; "fi" = Finnish; "fr" = French; "de" = German; "el" = Greek; "he" = Hebrew;" hi" = Hindi; "hu" = Hungarian; "it" = Italian; "ja" = Japanese; "ko" = Korean; "mr" = Marathi; "fa" = Persian; "pl" = Polish; "pt-br" = Portuguese-Brazilian; "ro" = Romanian; "ru" = Russian; "es" = Spanish; "sk" = Slovak; "sv" = Swedish; "zh" = Chinese; "th" = Thai; "uk" = Ukrainian.

The *n_words* and *n_chars* parameters control the granularity of summarization by words and characters, respectively, where setting -1 processes the entire text, trying to separate it into groups of sentences. The arguments *lowercase*, *rmv_accent*, *rmv_special_chars*, *rmv_numbers*, *rmv_custom_words,* and *verbose* offer various preprocessing capabilities, such as converting text to lowercase, removing accents, special characters, numbers, exclusion of specified terms, and providing execution updates. More details about text normalization are given in Sect. 2.1.

With the "smr" instance, it is possible to use the following extractive techniques:

(a) TextRank

It is a graph-based ranking algorithm for keyword and sentence extraction and an adaptation of the PageRank algorithm for web page ranking (Page et al., 1998). TextRank summarization involves text preprocessing, constructing a similarity matrix, creating a sentence graph, applying the PageRank algorithm, selecting top-ranked sentences, and generating the final summary. The *text_1* is preprocessed and tokenized by removing unnecessary elements. A similarity matrix captures semantic relationships between sentences. A graph is created with sentences as nodes and similarity scores as edges. PageRank assigns importance scores based on sentence connectivity. Top-ranked sentences are selected based on scores and concatenated to form the summary.

Figure 2 demonstrates the utilization of the TextRank function from the "smr" instance:

The argument *iteration* specifies the number of iterations for the algorithm, affecting the convergence of ranking scores. A higher number of iterations can lead to a more accurate but computationally intensive ranking. Argument *D* is a damping factor that adjusts the likelihood of jumping to a randomly selected sentence in the algorithm. A typical value of 0.85 suggests a 15% chance of random jumps, which

```
# TextRank
summary = smr.summ_text_rank(iteration = 1000, D = 0.85, model = 'all-MiniLM-L6-v2')
smr.show_summary(summary, n = 3)
```

Fig. 2 TextRank

helps distribute ranking scores more evenly across the set of sentences. The argument *model* selects the sentence embedding model. Here, *'all-MiniLM-L6-v2'* is chosen for transforming sentences into embeddings, which are then used to compute similarities. This specific model is part of the Sentence Transformers library, which generates high-quality embeddings for semantic similarity tasks. These embeddings capture the semantic essence of each sentence in a high-dimensional vector space. The *smr.show_summary(summary, n = 3)* command subsequently displays the top three sentences that most effectively encapsulate the essence of the text.

(b) LexRank

Like TextRank, it is also a graph-based algorithm, but it uses cosine similarity between TF-IDF (Term Frequency—Inverse Document Frequency) vectors of sentences to determine similarity. The LexRank score of a sentence is then calculated based on the average similarity of that sentence with all others, with the highest-ranked sentences selected for the summary. LexRank extractive summarization involves preprocessing the text, and a similarity matrix is created based on cosine similarity between sentences, forming a graph where sentences are nodes and edges represent similarity scores. The PageRank algorithm then assigns sentence importance scores based on connectivity and similarity. Top-ranked sentences are selected for the summary, with the number determined by the desired length. Finally, chosen sentences are concatenated to generate the summary in their original order or rearranged for coherence.

Figure 3 demonstrates the utilization of the LexRank function from the "smr" instance:

LexRank and TextRank are similar in their algorithmic approach to summarization, leveraging the concept of sentence importance within a document. However, LexRank distinguishes itself from TextRank by eschewing sentence embeddings, favoring a TF-IDF matrix for representing sentences. Each vector denotes a sentence in this vector space model, with components reflecting the TF-IDF scores of terms therein. This method emphasizes the distinctive elements of each sentence by weighting terms based on their frequency in a sentence against their commonality across the document.

```
# LexRank
summary = smr. summ_lex_rank (iteration = 1000, D = 0.85)
smr.show_summary(summary, n = 3)
```

Fig. 3 LexRank

```
# LSA
summary = smr.summ_ext_LSA(embeddings = False, model = 'all-MiniLM-L6-v2')
smr.show_summary(summary, n = 3)
```

Fig. 4 LSA

(c) LSA

Instead of working directly with the sentences, LSA works with the concepts or topics beneath the text. By applying singular value decomposition (SVD) to the DTM (Document-Term Matrix) or a word embedding matrix, LSA identifies clusters of semantically related terms, and sentences relevant to those topics are extracted for the summary. LSA extractive summarization involves preprocessing the text, creating a term-document matrix using SVD, calculating sentence similarities, constructing a weighted graph, applying graph algorithms like PageRank to assign importance scores, selecting top-ranked sentences as the summary, and concatenating them for the final summary.

Figure 4 demonstrates the utilization of the LSA function from the "smr" instance:

The argument *embeddings* dictate whether the algorithm should use sentence embeddings (True) or a TF-IDF matrix (False) for sentence representation. Setting this to False uses the TF-IDF matrix, aligning with traditional LSA practices that rely on this matrix's singular value decomposition (SVD) to identify patterns and relationships between terms and sentences. The argument *model* specifies the sentence embedding model to be used.

(d) KL Sum

It operates on the principle of minimizing the Kullback-Leibler divergence between the probability distribution of words in the document and the summary. It iteratively adds sentences to the summary so that the Kullback-Leibler divergence is minimized with each addition. KL Sum extractive summarization involves preprocessing the text, and a term-frequency (TF) matrix is then created to capture term importance in each sentence. Then, Kullback-Leibler divergence is computed using the matrix to determine the importance of the term relative to the document. Based on the divergence scores, top-ranked sentences are selected as the summary. The number of chosen sentences can be predetermined or dynamic. Finally, the selected sentences are concatenated to generate the summary, with the option to rearrange them for coherence.

Figure 5 demonstrates the utilization of the KL-Sum function from the "smr" instance:

The *n* argument determines how the algorithm selects a subset of sentences from the document to construct a summary that, ideally, minimally diverges from the original text's information distribution.

```
# KL Sum
summary = smr. summ_ext_KL(n = 3)
smr.show_summary(summary, n = 3)
```

Fig. 5 KL Sum

(e) BART

It is built on the Transformer framework and excels in various language generation tasks, including extractive text summarization, machine translation, and document completion. BART extractive summarization involves preprocessing the text, applying the BART model to generate contextualized representations, calculating attention scores for token importance, selecting top-ranked tokens as the summary, and concatenating them for the final summary.

Figure 6 demonstrates the utilization of the BART function from the "smr" instance:

The argument *model* specifies the pre-trained BART model to be used for summarization. The "facebook/bart-large-cnn" is explicitly designed for CNN/Daily Mail summarization tasks, making it adept at producing coherent and concise summaries from longer texts. The argument *max_len* defines the maximum length of the summary in terms of the number of tokens (words and punctuation). It ensures that the generated summary does not exceed 250 tokens, balancing detail and brevity to cater to the desired summary size.

(f) T5

An innovative model from Google Research, T5 handles text-to-text problems and solves many different NLP tasks. T5 summarization involves preprocessing the text, fine-tuning the T5 model on summarization datasets, feeding the document into the T5 model, generating a summary, and post-processing the summary if necessary. While it is primarily an abstractive summarization technique, it tends to include entire sentences from the original text, mainly when dealing with shorter texts. This behavior has led to its placement as an extractive technique in specific contexts.

Figure 7 demonstrates the utilization of the T5 function from the "smr" instance:

The argument *model* specifies the version of the T5 model to use. '*t5-base*' is a pre-trained model provided by Google that is balanced between size and performance and suitable for a wide range of text generation tasks, including summarization. The *min_len* sets the minimum length of the generated summary regarding the number of tokens, ensuring that the summary is not too brief and capturing essential information

```
# BART
summary = smr.summ_ext_bart(model = 'facebook/bart-large-cnn', max_len = 250)
print(summary)
```

Fig. 6 BART

```
# T5
summary = smr.summ_ext_t5(model = ' t5-base', min_len = 30, max_len = 250,
model_max_len = 512, verbose = False)
print(summary)
```

Fig. 7 T5

from the input text. The *max_len* determines the maximum length of the summary, also in tokens, preventing the summary from being overly lengthy and maintaining conciseness. The *model_max_length* configures the maximum number of tokens the model considers from the input text. Texts longer than the *model_max_length* threshold are truncated to the specified size. Finally, the argument *verbose* is a boolean flag that, when set to True, triggers the function to print the generated summary, as this can be useful for immediate feedback during development or debugging.

With the 'smr' instance, it is possible to use the following abstractive techniques:

(a) PEGASUS

To perform PEGASUS abstractive summarization, the model first undergoes pre-training on a large text corpus, with specific high-importance sentences masked for the model to predict. These sentences are selected based on their relevance to the overall document context, determined by an importance score derived from their ROUGE score. The model is then fine-tuned on a summarization-specific dataset, learning to encode entire documents into a latent space and generate summaries from this encoding. Once fine-tuned, PEGASUS can perform abstractive summarization on unseen text, generating new, concise sentences that encapsulate the key points from the original document.

Figure 8 demonstrates the utilization of the PEGASUS function from the "smr" instance:

The argument *model* specifies the PEGASUS model variant to be used. The "*google/pegasus-xsum*" model is particularly trained for summarization, aiming to generate concise and informative summaries. It's pre-trained to excel in summarization tasks by focusing on extracting salient information and generating a cohesive abstract. The *min_L* and *max_L* arguments define the minimum and maximum lengths of the generated summary regarding the number of tokens. This range ensures that the summary is neither too brief, missing crucial information, nor too lengthy, diluting the summary's effectiveness. The argument *verbose* is a boolean flag that, when set to True, triggers the function to print the generated summary, as this can be useful for immediate feedback during development or debugging.

```
# PEGASUS
summary = smr. summ_abst_pegasus (model = "google/pegasus-xsum ', min_L = 100,
max_L = 150)
print(summary)
```

Fig. 8 PEGASUS

```
# ChatGPT
summary = smr. summ_abst_chatgpt (api_key = 'your_api_key_here', query = 'make an abstratctive
summarization', model = 'gpt-4', max_tokens = 500, n = 1, temperature = 0.8)
print(summary)
```

Fig. 9 ChatGPT

(b) ChatGPT

It is a model OpenAI developed based on the GPT (Generative Pretrained Transformer) architecture. GPT models, particularly ChatGPT, are recognized for their impressive ability to generate human-like text. ChatGPT interprets the text and generates a condensed version that retains the key points and overall meaning. Summaries are produced word by word, with each new word generated based on previously generated context, creating entirely new sentences.

Figure 9 demonstrates the utilization of the ChatGPT function from the "smr" instance.

The argument *api_key* is a placeholder for the API key provided by OpenAI, which is necessary for authenticating and accessing the GPT services. The key enables the function to send requests to OpenAI's API to generate text completions. The *query* argument sets the task description prepended to the *text_1* to form the prompt. In this case, it instructs ChatGPT on the desired action to perform abstractive summarization. The *model* argument specifies the model variant of GPT to be used. The "gpt-4" model is known for its deep understanding and generation capabilities, making it suitable for complex tasks such as summarization. The *max_tokens* argument limits the length of the generated summary to the number of desired tokens. This argument helps ensure the output is concise and meets the desired length constraints. The *n* argument indicates that only one completion (or summary) should be generated per API call. The *temperature* argument controls the creativity or randomness of the generated summary. A temperature of 0.8 strikes a balance between being deterministic and allowing for some variability in the output, which can be beneficial for generating informative and engaging summaries. Finally, the argument *verbose* is a boolean flag that, when set to True, triggers the function to print the generated summary, as this can be useful for immediate feedback during development or debugging.

By combining extractive and abstractive techniques, pyAutoSummarizer offers users the flexibility and robustness needed for various text summarization tasks. The choice between extractive and abstractive methods, or a combination of both, allows users to customize the summarization process based on their specific requirements and the nature of the text.

Preprocessing Capabilities

Preprocessing text is an essential step in the text summarization process, and this step transforms raw text into a more digestible form, which allows subsequent algorithms to focus on the significant elements. pyAutoSummarizer offers an array of preprocessing capabilities, each contributing to creating more coherent, meaningful summaries.

- *Lowercase*: The text in different cases can be interpreted as distinct entities by algorithms, and to maintain uniformity and eliminate this issue, our library can convert all text to lowercase.
- *Removal of Accents*: Accents can introduce complexity and create unnecessary unique words in the text; our library can strip accents from characters, ensuring simplicity and consistency.
- *Removal of Special Characters*: Non-alphanumeric characters, such as punctuation and symbols, can add noise to text data. By eliminating these, our library keeps the focus on meaningful textual content.
- *Removal of Numbers*: Unless necessary for understanding, numbers often don't contribute to the meaning of the text and can be seen as noise; our library can remove these.
- *Removal of Custom Words*: Users can specify a list of custom words to be eliminated from the text. This feature is handy in specialized applications where specific frequent terms may not carry significant meaning.
- *Removal of Stopwords*: Some words often don't contribute to the overall meaning of a sentence that can be eliminated. Our library supports the removal of stopwords in many languages, facilitating multilingual text processing.
- *Sentence Segmentation*: The library can break down the text into sentences using punctuation, a specific number of characters, or a particular number of words. This capability allows users to adapt the granularity of summarization to their specific needs.

While preprocessing is an essential step in text analysis, the extent of preprocessing can vary depending on the problem. Some preprocessing steps may be optional, allowing users to tailor the level of normalization based on their preferences and requirements. In some instances, users may choose to perform minimal preprocessing, preserving the original characteristics of the text. This approach can be beneficial when specific nuances, such as capitalization or punctuation, play a crucial role in the analysis. Users can retain the text's unique characteristics relevant to their problem by exercising discretion over the preprocessing steps.

The most common steps are the Removal of Accents, simplifying the text by standardizing characters, and making further processing more uniform. Lowercasing ensures uniformity and prevents algorithms from misinterpreting the same words in different cases as distinct entities. Removal of Special Characters and Numbers is especially beneficial unless numbers or symbols carry specific meanings crucial to the summarization task. Removal of Stopwords can significantly declutter the text, making it easier for summarization algorithms to identify key information. Removal of Custom Words can remove frequent yet insignificant words for their specific application, further refining the focus of the summarization. This sequence is designed to refine the text progressively, eliminating irrelevant features and standardizing its format before segmenting it into analyzable units. However, users must consider their text's specific requirements and characteristics and summarization objectives, for instance, in cases where numerical data or special characters are significant for understanding.

On the other hand, there are scenarios where extensive preprocessing is necessary to enhance the quality and consistency of the data. Steps such as lowercasing, removing accents, eliminating special characters, and removing numbers can help reduce noise and standardize the text. Additionally, removing custom words and stopwords can refine the data by eliminating irrelevant or commonly occurring terms. The decision on which preprocessing steps to perform rests with the user. It is essential to balance preserving the text's inherent characteristics and applying appropriate normalization techniques to ensure the data is suitable for the desired analysis or task.

Evaluation Metrics

Another feature of pyAutoSummarizer is its ability to evaluate the quality of its generated summaries, which can be accessed by using well-known metrics:

(a) Rouge-N:

It refers to the overlap of N-grams between the system and reference summaries. An N-gram is a contiguous sequence of N items from a given sample of text or speech. For example, in Rouge-1, unigrams (single words) are considered; for Rouge-2, it's bigrams (two-word phrases), and so on. The evaluation begins by extracting n-grams of length N from the generated and reference summaries. Subsequently, the number of matching n-grams is counted, and a score is calculated, reflecting the ratio of matching n-grams to the total number in the reference summary, effectively measuring recall. The score, ranging from 0 to 1, indicates the overlap in content, with 1 implying a perfect match. However, ROUGE-N focuses on recall and doesn't account for precision or word order.

In Fig. 10, the argument *generated_summary* represents the summary produced by the selected summarization technique, which is the output summary you want to evaluate regarding its quality and relevance. The argument *reference_summary* represents the gold standard or the ideal summary against which the *generated_summary* is compared. It serves as the benchmark for evaluating the correctness and completeness of the generated summary. The argument *n* specifies the size of the n-grams to be used for the evaluation, and it allows the function to calculate the metric based on different granularities of text comparison.

```
# Rouge N
f1, precision, recall = smr.rouge_N(generated_summary = summary, reference_summary
= reference, n = 2)
```

Fig. 10 Rouge N

(b) Rouge-L

It uses the Longest Common Subsequence (LCS) between the system and reference summaries. The LCS is the lengthiest sequence of words appearing in both summaries in the same order, but not necessarily consecutively. The ROUGE-L score is calculated considering recall (the ratio of the LCS length to the reference summary length) and precision (the ratio of the LCS length to the generated summary length). The final ROUGE-L score often used is the F1 score, a harmonic mean of precision and recall, offering a balanced evaluation. The score ranges from 0 to 1; a higher ROUGE-L score indicates a higher similarity between the machine-generated and reference summaries in terms of both content and word order.

In Fig. 11, the argument *generated_summary* represents the summary produced by the selected summarization technique, which is the output summary you want to evaluate regarding its quality and relevance. The argument *reference_summary* represents the gold standard or the ideal summary against which the *generated_summary* is compared.

(c) Rouge-S

It considers skip-bigram co-occurrence statistics, where skip-bigrams are any pair of words in their sentence order, allowing for gaps. The core element is the identification of skip-bigrams (pairs of words in a sentence, not necessarily adjacent, with order considered) in both summaries. The ROUGE-S score is then calculated based on recall, i.e., the ratio of matching skip-bigrams in the generated summary to the total skip-bigrams in the reference summary. The score ranges from 0 to 1, with 1 indicating a perfect match. A higher ROUGE-S score indicates more significant similarity in content overlap and matching skip-bigram pairs, considering both content and limited word order.

In Fig. 12, the argument *generated_summary* represents the summary produced by the selected summarization technique, which is the output summary you want to evaluate regarding its quality and relevance. The argument *reference_summary* represents the gold standard or the ideal summary against which the *generated_summary* is compared. The argument *skip_distance* specifies the maximum distance

```
# Rouge L
f1, precision, recall = smr.rouge_L(generated_summary = summary, reference_summary
= reference)
```

Fig. 11 Rouge L

between words in a skip-bigram, defining how many words can be skipped over in the sequence to form a bigram. A skip distance of 4 allows for up to four words between the components of the bigram, enabling a flexible consideration of word order and proximity. This parameter adjusts the leniency in matching content between the summaries, with a larger skip distance accommodating a broader interpretation of similarity.

(d) BLEU

Initially developed to evaluate the quality of the machine-translated text, it's also used in the text summarization field. BLEU looks at the precision of N-grams, i.e., it checks how many N-grams in the generated summary are also present in the reference summary. The evaluation involves comparing n-grams of the generated summary to those in the reference summary, with precision calculated as the ratio of matching n-grams to the total n-grams in the generated summary. This step is done for different n values, typically from 1 to 4. A brevity penalty is included to avoid rewarding overly verbose summaries, reducing the score if the generated summary is shorter than the reference summary. The final BLEU score is the geometric mean of the n-gram precision scores adjusted by the brevity penalty, ranging from 0 to 1. Higher BLEU scores indicate a better quality of summary.

In Fig. 13, the argument *generated_summary* represents the summary produced by the selected summarization technique, which is the output summary you want to evaluate regarding its quality and relevance. The argument *reference_summary* represents the gold standard or the ideal summary against which the *generated_summary* is compared. The *n* argument specifies the maximum order of n-grams for the BLEU score calculation. By default, $n = 4$ is used, meaning that the metric considers unigrams ($n = 1$), bigrams ($n = 2$), trigrams ($n = 3$), and 4-grams ($n = 4$) to evaluate the precision of matching sequences of up to four words in the generated summary relative to the reference.

(e) METEOR

It aligns words in the generated and reference summaries, counting exact matches and similarities in roots, synonyms, and paraphrases. From this alignment, precision and

```
# Rouge S
f1, precision, recall = smr.rouge_S(generated_summary = summary, reference_summary
= reference, skip_distance = 2)
```

Fig. 12 Rouge S

```
# BLEU
score = smr.bleu(generated_summary = summary, reference_summary = reference, n =
4)
```

Fig. 13 BLEU

```
# METEOR
score = smr.meteor(generated_summary = summary, reference_summary = reference)
```

Fig. 14 METEOR

recall are calculated. The METEOR score is computed as the F1 score, the harmonic mean of precision and recall. A penalty is applied for word order differences and disjoint word sequences to avoid rewarding loosely related word lists. The score ranges from 0 to 1, with 1 indicating a perfect match. A higher METEOR score suggests a higher quality summary, accounting for both content overlap and semantic similarity.

In Fig. 14, the argument *generated_summary* represents the summary produced by the selected summarization technique, which is the output summary you want to evaluate regarding its quality and relevance. The argument *reference_summary* represents the gold standard or the ideal summary against which the *generated_summary* is compared. The *alpha* argument controls the F-mean calculation's balance between precision and recall, with a default value emphasizing precision slightly more than recall. The *beta* argument determines the sensitivity of the METEOR score to chunking penalties. A higher beta value makes the score more sensitive to the number and size of unmatched chunks between the generated and reference texts.

Integrating these metrics into the pyAutoSummarizer library empowers the system to evaluate and enhance the produced summaries. These evaluation metrics provide users with tangible evidence of the high quality of the generated summaries.

3 pyAutoSummarizer Text Summarization Example

Before exploring the specifics of our case study, we wish to highlight the inherent limitations of experimenting on a single instance. We fully recognize that the metrics and outcomes derived from this isolated example may not provide statistically comprehensive insights applicable to a broader range of scenarios. As such, the results and comparative analysis of algorithms presented in Table 3 should be interpreted with caution, bearing in mind the specific context and constraints of the case. This study aims not to ascertain one algorithm's superiority over others definitively but to illustrate potential applications and implications within a narrowly defined scenario. We believe, however, that even within its limited scope, this case study offers valuable insights and serves as a preliminary exploration that underscores the need for further, more extensive research. We encourage readers to view the findings as a springboard for broader investigation rather than conclusive evidence of algorithmic performance across the diverse spectrum of NLP applications.

To showcase the capabilities of pyAutoSummarizer, we will explore an excerpt adapted from Isaac Asimov's seminal work—Run Around—published in 1950. This excerpt introduces, for the first time, the three fundamental laws of robotics. These laws serve as the guiding principles that govern the behavior and interactions of intelligent machines. This experiment is available at: https://bit.ly/42lDAfR

Adapted excerpt from Asimov (1950):

In robotics, three fundamental laws govern robots' behavior and ethical considerations. The First Law dictates that a robot is prohibited from causing harm to a human being, either directly or by failing to take necessary action to prevent harm. The First Law upholds the value of human safety and prioritizes the well-being of individuals above all else. The Second Law establishes that a robot must adhere to the commands and instructions provided by humans unless those orders would contradict the First Law. Therefore the Second Law emphasizes the importance of human authority and ensures that robots operate within the bounds of ethical guidelines. Lastly, the Third Law mandates that a robot must safeguard its existence as long as self-preservation does not clash with the First or Second Law. This self-protective instinct helps ensure the robot's longevity and continued functionality without compromising human safety or violating human commands. Together, these three laws create a framework that governs the behavior of robots, aiming to balance the protection of human life, obedience to human authority, and preservation of the robot's existence.

A reference summary provides a point of reference to objectively assess the quality of the generated summaries and validate the effectiveness of the summarization techniques implemented in the library.

Reference Summary:

The First Law dictates that a robot may not injure a human being or, through inaction, allow a human being to come to harm. The Second Law establishes that a robot must obey the orders given to it by human beings except where such orders would conflict with the First Law. Lastly, the Third Law mandates that a robot must protect its existence as long as such protection does not conflict with the First or Second Law.

This demonstration will highlight pyAutoSummarizer's ability to condense complex texts while retaining the core ideas and key details, providing a glimpse into the summarization capabilities of the library.

Text Normalization

A series of normalization steps are applied to prepare the adapted excerpt for further analysis. These steps include removing stop words, converting the text to lowercase, and removing accents. In this case, there is no need to remove numbers or eliminate special characters or custom words from the text. Hence, the normalized excerpt, where each sentence is separated by punctuation, is obtained in Table 1:

Table 1 Normalized sentences

ID	Original sentences	Normalized sentences
1	In robotics, three fundamental laws govern robots' behavior and ethical considerations.	Robotics fundamental laws govern robots' behavior ethical considerations.
2	The first law dictates that a robot is prohibited from causing harm to a human being, either directly or by failing to take necessary action to prevent harm.	Law dictates robot prohibited causing harm human directly failing action prevent harm.
3	The first law upholds the value of human safety and prioritizes the Well-being of individuals above all else.	Law upholds human safety prioritizes individuals.
4	The second law establishes that a robot must adhere to the commands and instructions provided by humans unless those orders would contradict the first law.	Law establishes robot adhere commands instructions provided humans orders contradict law.
5	Therefore the second law emphasizes the importance of human authority and ensures that robots operate within the bounds of ethical guidelines.	Law emphasizes importance human authority ensures robots operate bounds ethical guidelines.
6	Lastly, the third law mandates that a robot must safeguard its existence as long as self-preservation does not clash with the first or second law.	Lastly law mandates robot safeguard existence long preservation clash law.
7	This self-protective instinct helps ensure the robot's longevity and continued functionality without compromising human safety or violating human commands.	Protective instinct helps ensure robot's longevity continued functionality compromising human safety violating human commands.
8	Together, these three laws create a framework that governs the behavior of robots, aiming to balance the protection of human life, obedience to human authority, and preservation of the robot's existence.	Laws create framework governs behavior robots aiming balance protection human life obedience human authority preservation robot's existence.

Generated Summaries

Table 2 presents the summaries generated by various models, including TextRank, LexRank, LSA (with Embeddings and TF-IDF), KL Sum, BART, T5, PEGASUS, and ChatGPT.

Upon analyzing Table 2 alongside the reference summary, it's clear that various summarization models interpret and distill the same source text differently. The TextRank and LexRank models have produced nearly identical summaries, highlighting the framework of laws and the balance they aim to create, emphasizing the Second Law and human authority. The LSA (Embeddings) and LSA (TF-IDF) summaries focus on the individual laws, each taking a slightly different approach. The former emphasizes the First and Third Laws, while the latter focuses on the First and Second Laws. The KL Sum model's summary is quite similar to the LSA(TF-IDF), but it also includes the Third Law. BART provides a comprehensive summary

Table 2 Generated summaries

ID	Generated summary
TextRank	Together, these three laws create a framework that governs the behavior of robots, aiming to balance the protection of human life, obedience to human authority, and preservation of the robot's existence. Therefore the second law emphasizes the importance of human authority and ensures that robots operate within the bounds of ethical guidelines. The second law establishes that a robot must adhere to the commands and instructions provided by humans unless those orders would contradict the first law.
LexRank	Together, these three laws create a framework that governs the behavior of robots, aiming to balance the protection of human life, obedience to human authority, and preservation of the robot's existence. Therefore the second law emphasizes the importance of human authority and ensures that robots operate within the bounds of ethical guidelines. This self-protective instinct helps ensure the robot's longevity and continued functionality without compromising human safety or violating human commands.
LSA (Embeddings)	The first law upholds the value of human safety and prioritizes the Well-being of individuals above all else. This self-protective instinct helps ensure the robot's longevity and continued functionality without compromising human safety or violating human commands. Lastly, the third law mandates that a robot must safeguard its existence as long as self-preservation does not clash with the first or second law.
LSA(TF-IDF)	In robotics, three fundamental laws govern robots' behavior and ethical considerations. The first law dictates that a robot is prohibited from causing harm to a human being, either directly or by failing to take necessary action to prevent harm. The second law establishes that a robot must adhere to the commands and instructions provided by humans unless those orders would contradict the first law.
KL sum	In robotics, three fundamental laws govern robots' behavior and ethical considerations. The second law establishes that a robot must adhere to the commands and instructions provided by humans unless those orders would contradict the first law. Lastly, the third law mandates that a robot must safeguard its existence as long as self-preservation does not clash with the first or second law.
BART	In robotics, three fundamental laws govern robots' behavior and ethical considerations. The first law upholds the value of human safety and prioritizes the Well-being of individuals. The second law emphasizes the importance of human authority and ensures that robots operate within the bounds of ethical guidelines. The third law mandates that a robot must safeguard its existence as long as self-preservation does not clash with the first or second law.
T5	Three laws govern robots' behavior and ethical considerations. First law prohibits robot from causing harm to a human being, second law emphasizes human authority, third law mandates self-preservation to ensure robot's longevity and continued functionality. Robot must obey human orders and obey commands provided by humans unless those orders contradict the first law.

(continued)

Table 2 (continued)

ID	Generated summary
PEGASUS	In our series of letters from African journalists, film-maker and columnist Ahmed Rashid looks at the importance of self-preservation in the world of robotics, and how humans can play a role in ensuring that robots do not harm themselves or others. Together, these three laws create a framework that the behavior of robots, aiming to balance the protection of human life, obedience to human authority, and preservation of the robot's existence., and the preservation of the robot's existence., and the first and second laws.
ChatGPT	This article explores the three fundamental laws governing robotics which prioritize the safety of humans, ensure obedience to human commands, and protect the robot's existence. These laws create a framework of ethical considerations to ensure robots operate within the bounds of safety and responsibility.

covering all three laws and their specific functions, aligning closely with the reference summary. The T5 model's summary is the most condensed, representing the main ideas but missing some details. PEGASUS, however, includes an unrelated statement about African journalists in its summary, unrelated to the original text about robotic laws; this indicates that PEGASUS was hallucinating content. This tendency toward hallucination can be attributed to the model's training process and the inherent complexity of language understanding. During pre-training, these models are trained on a wide range of data and may pick up patterns or themes they inject into their generated content, even when inappropriate. ChatGPT generates a clear, concise, and general summary that does not delve into the details of the individual laws but accurately presents the overarching principles. Comparing these with the reference summary, BART and LSA(TF-IDF) provide the closest matches in terms of content coverage, but the wording varies. All models capture the text's central theme about the three governing laws of robotics, but there are differences in detail and emphasis.

Evaluation Metrics

Table 3 presents the evaluation metrics for various summarization models. Each model's performance is measured through five evaluation metrics: Rouge-N, Rouge-L, Rouge-S, BLEU, and METEOR.

Analyzing Table 3, we can perceive that TextRank and LexRank models show moderate performance, with TextRank slightly outperforming LexRank on most metrics. TextRank scores 0.30 on Rouge-N, 0.41 on Rouge-L, 0.24 on Rouge-S, 0.25 on BLEU, and 0.61 on METEOR. In comparison, LexRank scores slightly lower in all categories except for METEOR, which scores 0.52. LSA (Embeddings) outperforms these models significantly with scores of 0.50, 0.58, 0.43, 0.42, and 0.63 on Rouge-N, Rouge-L, Rouge-S, BLEU, and METEOR, respectively. However, LSA(TF-IDF) struggles with scores of 0.00, 0.18, 0.03, 0.19, and 0.33 on the same metrics. KL Sum

Table 3 Metrics

	Rouge N (F1 score)	Rouge L (F1 score)	Rouge S (F1 score)	BLEU	METEOR
TextRank	0.30	0.41	0.24	0.25	0.61
LexRank	0.28	0.30	0.20	0.21	0.52
LSA (Embeddings)	0.50	0.58	0.43	0.42	0.63
LSA(TF-IDF)	0.00	0.18	0.03	0.19	0.33
KL sum	0.31	0.37	0.30	0.28	0.43
BART	0.77	0.81	0.66	0.64	0.93
T5	0.07	0.27	0.08	0.13	0.38
PEGASUS	0.00	0.09	0.00	0.13	0.17
ChatGPT	0.00	0.12	0.02	0.11	0.11

presents better results than LSA(TF-IDF) and LexRank but underperforms compared to TextRank and LSA (Embeddings), with scores of 0.31, 0.37, 0.30, 0.28, and 0.43. BART outperforms all the other models by a substantial margin, achieving impressive scores of 0.77, 0.81, 0.66, 0.64, and 0.93. T5, PEGASUS, and ChatGPT models show limited performance compared to others. T5 scores 0.07, 0.27, 0.08, 0.13, and 0.38; PEGASUS has the lowest scores with 0.00, 0.09, 0.00, 0.13, and 0.17; and ChatGPT scores 0.00, 0.12, 0.02, 0.11, and 0.11.

The performance of abstractive models may appear inferior when evaluated solely based on metrics that compare the generated summaries to a reference summary, and this is because these models aim to create novel sentences that capture the essence of the source text rather than directly copying phrases, a process known as abstraction. While the reference comparison approach works well for extractive models (which essentially pick out key sentences or phrases from the source text), it might not fairly assess the abstractive models. The reference summary may contain specific phrasings or structures that the abstractive models do not replicate, leading to lower scores even if the generated summary captures the same information accurately and coherently. Therefore, abstractive models' performance could be more accurately assessed if the evaluation metrics considered semantic similarity or meaning preservation rather than relying heavily on the exact phrase or structure matching. Furthermore, human evaluation could also be beneficial, as humans are generally better at understanding whether a summary, although phrased differently, captures the intended meaning of the original text.

Still, regarding evaluating abstractive models, two particularly promising metrics have been developed that more accurately capture their capabilities: BERTScore and MoverScore. Developed by Zhang et al. (2020a, 2020b), BERTScore has demonstrated a strong correlation with human judgments concerning text quality, suggesting its effectiveness in evaluating the nuances of summarization. MoverScore, introduced by Zhao et al. (2019), excels in capturing more abstract textual relationships, offering a deeper analysis of the semantic layers within summaries. Although these metrics mark a significant advancement over traditional methods, it is acknowledged that further research is required to assess abstractive summaries fully. As the field develops and new metrics undergo empirical validation and gain consensus within the academic community, we are committed to integrating these innovations into pyAutoSummarizer.

It is pertinent to note that each algorithm's efficacy varies with the length and complexity of the input text. Extractive methods like TextRank and LexRank are generally less constrained by text length, efficiently handling long and short texts by identifying key sentences based on their structural and semantic importance. However, these methods might not always capture the nuanced essence of very long texts with equal effectiveness. In contrast, abstractive models, built on advanced neural network architectures, are adept at summarizing longer paragraphs by generating cohesive and concise summaries. These models, however, require substantial computational resources, particularly for longer texts, which could impact their practical applicability in resource-constrained environments. LSA is effective across a range of text lengths but is particularly suited to texts where the extraction of underlying themes is more important than the precise reproduction of source text. Each technique, therefore, presents a trade-off between efficiency, accuracy, and computational demand, with the choice of the method depending on the specific requirements of the summarization task, including the length and complexity of the text to be summarized.

Various algorithms and models generally offer tailored solutions for text summarization tasks to suit distinct needs and content types. TextRank is exceptional for extractive summarization, pinpointing and drawing out key sentences central to the text's message, making it particularly suitable for academic works and detailed reports. LexRank builds on a similar premise but introduces eigenvector centrality to assess sentence importance within semantically similar texts, which is ideal for cohesive thematic documents like news articles. LSA examines deeper, extracting the core concepts or themes by reducing text complexity, which is invaluable for texts rich in diverse topics or intricate themes. KL Sum, utilizing Kullback-Leibler divergence, excels in refining texts laden with redundancy by selecting sentences that introduce novel information, a boon for evolving news or technical documentation. On the transformative front, BART crafts contextually pertinent summaries suitable for complex narratives, whereas T5 rephrases and condenses text, aiming for succinctness, making it apt for executive summaries or broad audience communications. PEGASUS stands out in abstractive summarization, producing human-like summaries perfect for feature articles or comprehensive reports. Lastly, ChatGPT's versatility in adapting to diverse text genres and styles allows for highly customized summaries, meeting a broad spectrum of summarization needs from content creation to educational materials, showcasing the adaptability and range of modern summarization tools.

4 Advantages and Limitations

As with any software library, pyAutoSummarizer has strengths and areas that could improve further. Here, we'll discuss the key advantages of the library, as well as some potential limitations. pyAutoSummarizer is characterized by a comprehensive range of summarization methods, offering extractive algorithms (including TextRank, LexRank, LSA, KL-Sum, and T5) and abstractive models (such as BART, PEGASUS, and ChatGPT). This wide array of methodologies allows users to choose the best-fitting approach based on their specific needs.

The library also integrates popular evaluation metrics such as Rouge-N, Rouge-L, Rouge-S, BLEU, and METEOR. Table 3 illustrates the utilization of these metrics, providing a convenient way for users to assess the quality of the generated summaries and refine their chosen techniques. The library is designed with a strong focus on usability, and simple pip installation, making it user-friendly for beginners and experienced NLP practitioners.

However, it does have some limitations. The library relies on external models (T5, BART, PEGASUS, and ChatGPT). These relatively large models can lead to increased resource demands, potentially posing a challenge in environments with limited computational resources. Additionally, it's important to note that models like ChatGPT, in particular, are part of the OpenAI GPT family and may come with usage limitations or associated costs, especially for commercial or production use. Users should be mindful of these considerations when utilizing the ChatGPT model in the pyAutoSummarizer library.

Another limitation is the variability in the quality of the summaries generated. Despite incorporating advanced algorithms and models to optimize summary quality, as shown in Table 3, discrepancies in performance can occur. It's important to remember that such inconsistency is a common challenge in automatic text summarization, not confined to the pyAutoSummarizer library alone.

5 Conclusion

The importance of text summarization in NLP is undisputed. Applications range from news aggregation and document summarization to customer feedback analysis. The ability to derive concise, meaningful summaries from vast textual data is vital for knowledge extraction and decision-making. The complexity of human language amplifies this challenge. In this context, pyAutoSummarizer, a comprehensive Python library, stands as a robust solution for automated text summarization.

Throughout this article, we've examined the various facets of pyAutoSummarizer, emphasizing its diverse extractive and abstractive summarization techniques. By incorporating advanced algorithms like TextRank, LexRank, LSA, and KL-Sum and sophisticated machine learning models like BART, T5, PEGASUS, and ChatGPT, pyAutoSummarizer equips users with the necessary tools to generate high-quality, contextually accurate summaries.

A noteworthy feature of pyAutoSummarizer is its integration of prominent evaluation metrics, including Rouge-N, Rouge-L, Rouge-S, BLEU, and METEOR. As illustrated in Table 3, these metrics enable users to assess their summarization outputs quantitatively. However, as we explored earlier, these metrics could exhibit a bias against abstractive methods, remarkably when evaluated against a singular reference summary. This limitation should be considered when interpreting the performance of abstractive models.

Despite limitations like the resource demands for large models and occurrences of hallucination (unrelated information generated by models, such as PEGASUS), the library remains valuable in the NLP toolkit due to its versatility, scalability, and user-friendliness. Furthermore, the evolving nature of open-source projects suggests that future updates could address these challenges.

In conclusion, pyAutoSummarizer stands as a significant asset, simplifying the complex task of text summarization. As the NLP field evolves, tools like pyAutoSummarizer will be crucial in transforming how we extract and interpret the ever-growing volumes of textual data in our digital age. As practitioners in the field, we eagerly anticipate the potential enhancements and new features this powerful library will introduce. pyAutoSummarizer is not just a summarization tool; it's a substantial stride forward in automating the understanding and interpretation of our expansive textual universe.

References

Asimov, I. (1950). *Run around. I, robot (the Isaac Asimov collection ed.)*. Doubleday.

Banerjee, S., & Lavie, A. (2005). METEOR: An automatic metric for MT evaluation with improved correlation with human judgments. In *Proceedings of Workshop on Intrinsic and Extrinsic Evaluation Measures for MT and/or Summarization at the 43rd Annual Meeting of the Association of Computational Linguistics (ACL-2005)*. https://aclanthology.org/W05-0909.

Collobert, R., Weston, J., Bottou, L., Karlen, M., Kavukcuoglu, K., & Kuksa, P. (2011). Natural language processing (almost) from scratch. *Journal of Machine Learning Research, 12*, 2493–2537. https://doi.org/10.5555/1953048.2078186

Edmundson, H. P. (1969). New methods in automatic extracting. *Journal of the ACM (JACM), 16*, 264–285. https://doi.org/10.1145/321510.321519

Erkan, G., & Radev, R. D. (2004). LexRank: Graph-based lexical centrality as salience in text summarization. *Journal of Artificial Intelligence Research, 22*, 457–479. https://doi.org/10.1613/jair.1523

Gambhir, M., & Gupta, V. (2017). Recent automatic text summarization techniques: A survey. *Artificial Intelligence Review, 47*, 1–66. https://doi.org/10.1007/s10462-016-9475-9

Ghag, K. V., & Shah, K. (2015). Comparative analysis of effect of stopwords removal on sentiment classification. In *Proceedings of the 2015 international conference on computer, communication and control (IC4)* (pp. 1–6). https://doi.org/10.1109/IC4.2015.7375527

Haghighi, A., & Vanderwende, L. (2009). Exploring content models for multi-document summarization. In *Proceedings of the 2009 human language technologies* (pp. 362–370). Boulder.

Hossain, E., Rana, R., Higgins, N., Soar, J., Barua, P. D., Pisani, A. R., & Turner, K. (2023). Natural language processing in electronic health records in relation to healthcare decision-making: A systematic review. *Computers in Biology and Medicine, 155*(106649). https://doi.org/10.1016/j.compbiomed.2023.106649

Lewis, M., Liu, Y., Goyal, N., Ghazvininejad, M., Mohamed, A., Levy, O., Stoyanov, V., & Zettlemoyer, L. (2019). *BART: Denoising sequence-to-sequence pre-training for natural language generation, translation, and comprehension.* https://doi.org/10.48550/arXiv.1910.13461

Lin, C.-Y., & Hovy, E. (2003). Automatic evaluation of summaries using N-gram co-occurrence statistics. In *Proceedings of the 2003 human language technology conference of the north American chapter of the Association for Computational Linguistics* (pp. 150–157) https://aclanthology.org/N03-1020.pdf

Lin, C-Y., & Och, F.J. (2004). Automatic evaluation of machine translation quality using longest common subsequence and skip-bigram statistics. In *Proceedings of the 42nd Annual Meeting of the Association for Computational Linguistics (ACL 2004).* https://doi.org/10.3115/1218955.1219032.

Luhn, H. P. (1958). A business intelligence system. *IBM Journal of Research and Development, 2*, 314–319. https://doi.org/10.1147/rd.24.0314

Mihalcea, R., & Tarau, P. (2004). TextRank: Bringing order into text. In *Proceedings of the 2004 conference on empirical methods in natural language processing* (pp. 404–411). Association for Computational Linguistics. https://aclanthology.org/W04-3252.pdf.

Miso, R. (2023). Sumy: An automatic text summarization tool (version 0.11.0) [software]. GitHub. https://github.com/miso-belica/sumy.

Mutlu, B., & Sezer, E. A. (2023). Enhanced sentence representation for extractive text summarization: Investigating the syntactic and semantic features and their contribution to sentence scoring. *Expert Systems with Applications, 227*(120302). https://doi.org/10.1016/j.eswa.2023.120302

OPENAI. (2023). *GPT-4 technical report*. https://doi.org/10.48550/arXiv.2303.08774

Page, L., Brin, S., Motwani, R., & Winograd, T. (1998). The PageRank citation ranking: Bringing order to the web. In *Proceedings of the 7th international world wide web conference* (pp. 161–172). Brisbane.

Papineni, K., Roukos, S., Ward, T., & Zhu, W. J. (2002). BLEU: A method for automatic evaluation of machine translation. In *ACL-2002: 40th Annual Meeting of the Association for Computational Linguistics* (pp. 311–318). CiteSeerX 10.1.1.19.9416. https://doi.org/10.3115/1073083.1073135.

Raffel, C., Shazeer, N., Roberts, A., Lee, K., Narang, S., Matena, M., Zhou, Y., Li, W., & Liu, L. J. (2020). *Exploring the limits of transfer learning with a unified text-to-text transformer*. https://doi.org/10.48550/arXiv.1910.10683

Řehůřek, R., & Sojka, P. (2011). Gensim-python framework for vector space modelling. *NLP Centre, Faculty of Informatics, Masaryk University, Brno, Czech Republic, 3*(2).

Steinberger, J., & Karel, J. (2004). Using latent semantic analysis in text summarization and summary evaluation. In *Proceedings of the 2004 International Conference on Information Systems Implementation and Modelling—ISIM*.

Zhang, J., Zhao, Y., Saleh, M., & Liu, P. J. (2020). *PEGASUS: Pre-training with extracted gap-sentences for abstractive summarization.* https://doi.org/10.48550/arXiv.1912.08777

Zhang, T., Kishore, V., Wu, F., Weinberger, K. Q., & Artzi, Y. (2020). *BERTScore: Evaluating text generation with BERT*. https://doi.org/10.48550/arXiv.1904.09675

Zhao, W., Peyrard, M., Liu, F., Gao, Y., Meyer, C. M., & Eger, S. (2019). *MoverScore: Text generation evaluating with contextualized embeddings and earth mover distance.* https://doi.org/10.48550/arXiv.1909.02622

Measuring Technological Convergence in Encryption Technologies with Proximity Indices: A Text Mining and Bibliometric Analysis Using OpenAlex

Alessandro Tavazzi, Dimitri Percia David, Julian Jang-Jaccard, and Alain Mermoud

1 Introduction

In an era characterized by a technological revolution, understanding the dynamics of technological evolution, convergence, and emergence has become crucial for advancing science and fostering economic innovation (Carley et al., 2018; Halaweh, 2013; Wang et al., 2019; Zhou et al., 2019). While emergent technologies continue to reshape global landscapes socially, economically, and intellectually (Porter et al., 2019), a substantial gap exists in the literature concerning a comprehensive quantitative measure for assessing technological convergence (Curran et al., 2010; Song et al., 2017a).

To address this critical gap, our study employs the integration of bibliometric indicators, such as collaboration, common keywords, and citations, to facilitate a multidimensional analysis of technological convergence. We leverage the OpenAlex database, a rich source of scholarly papers, to model the evolution of encryption technology from 2002 to 2022. Unlike previous studies that typically use the binary relationship between research and the concept attribution to a specific technology, we

A. Tavazzi (✉) · D. P. David · J. Jang-Jaccard · A. Mermoud
Cyber-Defence Campus, armasuisse Science and Technology, Lausanne, Switzerland
e-mail: tavazale@gmail.com

A. Tavazzi
Institute of Mathematics, Lausanne, Switzerland

D. P. David
Institute of Entrepreneurship & Management, University of Applied Sciences of Western Switzerland (HES-SO Valais-Wallis), Sierre, Switzerland

H. da Hora et al. (eds.), *Technology Mining*,
https://doi.org/10.1007/978-3-032-10849-4_3

leverage OpenAlex's technology attribution scores. This enhances the relational granularity between research papers and specific technological concepts, thus improving the accuracy of identifying technological convergence.

Additionally, we use random forests to generate time series of proximity indices to forecast technological trajectories. This approach has allowed us to identify a significant convergence between blockchain and public-key cryptography, aligning well with the trend in the growing practical application of public-key cryptography within blockchain ecosystems.

By identifying early stages of technological convergence, our study not only complements but also extends traditional patent analyses, providing better insight into emerging technological trends. This foresight can guide strategic investment and development in cybersecurity to strengthen defenses against evolving cyber threats.

The paper is organized as follows. In Sect. 2, we evaluate the strengths and weaknesses of existing literature. Section 3 outlines the data processing methodology, and Sect. 4 presents the details of our proposed model. The results and their interpretations are presented in Sect. 5. Finally, Sect. 6 discusses the limitations of our study, potential extensions, and offers concluding remarks.

2 Related Work

This section provides an overview of the current state of the art in technology convergence approaches. Additionally, we explore various measurements and models that researchers have used to identify technological convergence, particularly in the context of emerging technologies.

Technology Convergence Approaches

The approaches used to assess technological convergence are diverse, with many researchers relying on proximity indices between technologies. While patent-based approaches are common in the literature (Song et al., 2017a, 2017b; Lee, 2018; Kim, 2017; Curran et al., 2010), our focus shifts toward a bibliometric-based approach. These approaches often utilize metrics like citation counts, publication frequencies, and keyword connections (Chand & Bhatt, 2021). In addition, most existing research focuses on convergence after academic discovery but before market implementation (Curran et al., 2010). However, these studies often lack comprehensive early-stage scientific analysis, a gap we aim to address in our approach.

Emerging technologies have a significant presence in literature, with diverse definitions (Halaweh, 2013; Liu & Xia, 2015; Carley, 2018). We compare two main factors: one looks at socioeconomic factors like uncertainty and cost, while the other emphasizes scientific features such as novelty, growth, and community integration.

In this study, we choose the latter, as it aligns with our focus on early-stage scientific convergence.

In existing works, technology convergence is typically studied separately from emerging technologies, often utilizing completely different sets of approaches. However, in our study, we aim to identify technology convergence in emerging technologies by combining and utilizing methods and indicators that are often employed separately.

Technology Convergence Measurements

In this section, we review the state of the art in bibliometric indicators, such as keywords, citations, and collaborations, along with our critiques.

3 Keywords

Identifying common keywords across various technologies is considered a crucial measure of technological convergence. This approach, especially its variant focusing on dynamic networks formed based on shared keywords, has been central in several studies (Blouin et al., 2013; Chand Bhatt et al., 2021; Huang et al., n.d.; Kim & Sohn, 2020; Song et al., 2017b). Despite its usefulness, keyword analysis has limitations, including ambiguities from semantic similarities in common keywords and imprecision in individual keywords representing distinct concepts (Chen et al., 2016; Li et al., 2016). Nevertheless, our study assumes that a shared lexicon can be an important indicator of convergent objectives, implying technological convergence. This assumption aligns with the notion that a common vocabulary, such as keywords, typically signifies a shared understanding of interest in a certain technology (Cojocaru et al., 2012), thus widely accepted as an important measure (Blouin et al., 2013; Chand Bhatt et al., 2021; Huang et al., n.d.; Kim & Sohn, 2020; Song et al., 2017b).

Citations

Citation and co-citation analyses, serving as indicators of scholarly interdependence, are essential for understanding technological convergence (Curran et al., 2010; He et al., 2022; Klarin et al., 2021; Li et al., 2016).

However, these methods face challenges stemming from incomplete datasets and unreliable reference-listing algorithms (Curran et al., 2010). Moreover, McRoberts' influential critique emphasized the limitations of citation analysis in fully capturing academic interconnectedness, pointing to unacknowledged influences, citation biases, literature ignorance, and self-citation practices as contributing factors (MacRoberts & MacRoberts, 1996).

Despite these limitations, we make the assumption that citation metrics, while offering only a partial perspective, provide insights into knowledge exchange across various domains, indicating potential convergence (Curran et al., 2010; Klarin et al., 2021). This perspective is supported by insights from Br¨öring, who suggests that interdisciplinary citations often signal the onset of scientific convergence, potentially leading to collaborative research ventures (Curran et al., 2010).

Collaborations

In addition to keywords and citations, the research community recognizes other indicators of technological convergence. One such metric involves assessing the extent to which researchers contribute to multiple technological fields. We argue that this aspect of research collaborations plays a crucial role in driving scientific and technological advancements (Curran et al., 2010; Liu & Xia, 2015). From an organizational theory standpoint, collaborations signify the preliminary stages of technological convergence (Curran et al., 2010; Liu & Xia, 2015). As articulated by Br¨öring, an increase in collaborative efforts typically precedes the merging of distinct disciplines (Curran et al., 2010). Therefore, analyses focusing on co-authorships and shared researcher pools hold value in predicting technological convergence.

Technology Convergence Models

This section explores graph-based and forecasting models used to identify and predict clusters of converging technologies from among other common technologies.

Graph-Based Models

Graph-based strategies provide an alternative, allowing for an examination of technological convergence from a macro perspective, rather than relying solely on isolated pairwise analysis (Breitzman & Thomas, 2015; Chand Bhatt et al., 2021; Huang et al., n.d.; Kim et al., 2019; Kumar et al., 2020; Lu¨ & Zhou, 2011). Cluster analyses and hypergraphs facilitate this process, aiding in the identification and prediction of converging technology clusters (Breitzman & Thomas, 2015; Chand Bhatt et al.,

2021; Huang et al., n.d.; Kumar et al., 2020; Lu¨& Zhou, 2011). This approach recognizes that technological convergence generally occurs within groups, highlighting the limitations of pairwise analyses in a field characterized by interconnected networks (Breitzman & Thomas, 2015; Schoen et al., 2012). However, the broader perspective of graph-based methods may obscure subtle interactions at the micro-level, potentially oversimplifying or misrepresenting complex relationships and leading to information loss or biased interpretations.

Forecasting Models

Forecasting models for technological convergence leverage a diverse range of algorithms and methodologies. These models incorporate graph-based clustering algorithms such as Spectral Louvain Modularity (SLM) and the Louvain method, Design Structure Matrix (DSM) tools, random forests, and topological clustering to analyze the connections between technological evolutions and emergences (Huang et al., n.d.; Kajikawa & Takeda, 2008; Kim & Lee, 2017; Kim & Sohn, 2020; Kumar et al., 2020; Lee et al., 2018). In addition, predictive models utilize AutoRegressive Integrated Moving Average (ARIMA), neural networks, and exponential smoothing techniques to anticipate technological trends (Kim & Lee, 2017; Kim & Sohn, 2020; Lee et al., 2018; Mitrea et al., 2009).

4 Data

This study utilizes data from OpenAlex, a comprehensive repository of research papers, authors, and institutions, which was created by drawing aspirations of the ancient Library of Alexandria (Priem et al., 2205). OpenAlex employs an automated system that assigns a set of Wikidata concepts to each paper from a pool of 65,000 concepts. This is accomplished through a multi-class deep learning classifier trained on the Microsoft Academic Graph. Each concept is assigned a score between 0 and 1, indicating the paper's relevance to that particular concept (Priem et al., 2205).

Data Extraction and Processing

Data relevant to the evolution of encryption technologies was extracted and processed into time series for analysis and forecasting. The focus was on scientific concepts related to 36 encryption technologies identified by experts through the Delphi method (Brown et al., 1968). Among the 36 technologies, only 25 were found in OpenAlex, forming the basis of our analysis. Papers not relevant to these 25 technologies were assigned a zero score. It was observed that over 90% of the papers had a zero-attribution score, affirming the importance of the scoring system (refer to Fig. 5 in the appendix).

Data Refinement

Due to the incomplete data in OpenAlex, papers without references, constituting half of the dataset, were removed to prevent bias. The distribution of these papers across technologies was uniform, ruling out any potential data skew (see Fig. 6 in the appendix). Additionally, we excluded another 5% of the remaining papers that were not linked to any concepts. To address anomalies like the overrepresentation of papers published in January, we corrected them by evenly redistributing throughout the year. Duplicate entries were resolved by keeping the version with the most comprehensive information to the set of papers under consideration.

Data Enhancement

Keywords were assigned to each paper using KeyBert,[1] a model that identifies keywords based on their similarity to the text, utilizing "cosine similarity" to indicate keyword relevance. Although a substantial portion of keywords was unique, common keywords played a crucial role in our analysis (see Fig. 2 in the appendix).

Author Influence

We calculated the h-index for authors in the encryption field from 2002 to 2022, generating both incremental and non-incremental indices on both a monthly and yearly basis. The incremental indices, offering a cumulative perspective of an author's impact, were considered a more accurate representation of an author's significance in the field (refer to Figs. 3 and 4 in the appendix).

Data Consolidation

The complete dataset included different elements like h-indices, references, and keywords. These were organized into a detailed data frame with columns such as "paper," "keyword," "cosine similarity," "title," "publication date," "abstract," "year," "month," "author," "referenced works," "concepts," "score concepts," "yearly H index not incremental," "yearly H index incremental," "monthly H index incremental," and "monthly H index not incremental."

5 Method

This section outlines the specifics of our proposed method, introducing novel mechanisms for calculating proximity indices using keywords, collaborations, and citations. Additionally, we delve into the details of how we fitted the data through interpolation to enhance the clustering and forecasting capabilities of the proposed method.

---

[1] https://maartengr.github.io/KeyBERT/api/keybert.html.

Proximity Indices

We aim to understand if certain indices capture specific stages of convergence more effectively than others. Additionally, we explore whether these indicators consistently align or may contradict each other in certain cases. Our hypothesis is that regardless of the indicator used to model technological convergence, they will generally reveal convergence between two technologies if it exists. However, this convergence may manifest differently or at different times. To investigate this, we create multiple indicators of technological convergence based on common keywords, citations, and collaboration between technologies. We assign papers that are attributed to technologies with a score between 0 and 1. Similarly, computed keywords are assigned to papers with a "cosine similarity" score between 0 and 1. To capture the importance of each author in the field of encryption technologies, we compute an artificial h-index specific.

For each indicator, our approach involves incorporating all relevant information for each month of each year, assigning appropriate weights to the variables, and constructing a time series. For example, in the case of keywords, we compute all common keywords between two technologies for each month of each year. We then calculate the average number of occurrences of each keyword in papers related to these two technologies during that month. This value is multiplied by the average cosine similarity and the average score of attribution to the pairs of technologies of the papers where each keyword occurred. These weights reflect the importance of specific keywords and the pair of technologies at a given moment in time. Similar computations are performed for other indices.

It's important to note that the indices presented in this paper draw inspiration from literature, where common keywords, collaboration, and citations are commonly used. However, the explicit form of the indices is novel, as they directly leverage a unique database attributing every paper to concepts with a score between 0 and 1.

Furthermore, we opt not to normalize the indices. Common normalization factors are either volatile, introducing bias, or yield very small indices, resulting in flat curves that hinder the identification of trends in technological convergence. The non-normalized time series of all indicators are presented on a single plot for all pairs of technologies.

Each index discussed below is computed for a specific month from 2002 to 2022. However, for simplicity, we denote the variables used in the computation of the indices without explicitly specifying their dependence on a month, even though these variables are inherently tied to a specific month.

Keywords Proximity Index

Let $t1$ and $t2$ be two technologies. Let $\mathrm{K}_{\mathrm{t1-t2}}$ denote a keyword: the keyword of some papers related to $t1$ and some papers related to $t2$ published during a specific month. Let $\mathrm{C}^{\mathrm{K}}_{\mathrm{t1-t2}}$ be the average cosine similarity of all the times this keyword occurred as a keyword of a paper related to $t1$ or $t2$ for a specific month. This measures how significant the keyword is for the papers related to $t1$ or $t2$ for a specific month. Let

A^{K}_{t1-t2} be the average attribution score to $t1$ and $t2$ of all papers related to $t1$ or $t2$ published during a specific month, with K as a keyword. This represents how close the papers having K as a keyword are to the technologies $t1$ and $t2$ during a specific month. Let N^{K}_{t1-t2} be the average of times K occurs in papers related to $t1$ and to $t2$, which were published during a specific month. For a given month m, we define the index of proximity based on keywords between two technologies $t1$ and $t2$ as follows:

$$IK^{m}_{t1-t2} := \sum_{K_{t1-t2}} N^{K}_{t1-t2} * C^{K}_{t1-t2} * A^{K}_{t1-t2}$$

This index aims to measure how significant a keyword is for the proximity between technology $t1$ and $t2$. The idea is, for each common keyword between both technologies, to multiply $C^{K}_{t1-t2}*A^{K}_{t1-t2}$, which works as a coefficient of importance for the keyword, with N^{K}_{t1-t2} the number of times the keyword appears in papers related to both technologies. Then, we take the given sum over all common keywords to obtain a global degree of proximity. As a result, this index evaluates the proximity of technologies based on weighted common keywords present in both technologies.

Citations Proximity Index

Let $t1$ and $t2$ be two technologies. Let C_{t1-t2} denote a paper related to the technology $t1$, which cites a paper related to the technology $t2$. Let $A^{C_{t1-t2}}_{t1}$ be the score of attribution to the technology $t1$ of the paper C_{t1-t2} and $A^{C_{t1-t2}}_{t2}$ be the score of attribution to the technology $t2$ of the paper related to the technology $t2$ cited by the paper C_{t1-t2}.

For a given month m, we define the two indices of proximity based on citations between two technologies, from $t1$ to $t2$ and from $t2$ to $t1$ as follows:

$$IC^{m}_{t1-t2} := \sum_{C_{t1-t2}} \left(A^{C_{t1-t2}}_{t1} + A^{C_{t1-t2}}_{t2} \right) * \frac{1}{2}$$

$$IC^{m}_{t2-t1} := \sum_{C_{t2-t1}} \left(A^{C_{t2-t1}}_{t2} + A^{C_{t2-t1}}_{t1} \right) * \frac{1}{2}$$

These indices measure the flow of citations from one technology to another during a given month. Furthermore, it assesses the importance of each citation for the connection between the two given technologies $t1$ and $t2$. As a result, these indices evaluate the proximity of technologies based on citations.

Collaboration Proximity Index

Let $t1$ and $t2$ be two technologies. Let A_{t1-t2} denote an author who is the author of some papers related to $t1$ and some papers related to $t2$ published during a specific month. For an author A_{t1-t2}, let $H^{A_{t1-t2}}_{n}$ be its monthly not incremental h-index and $H^{A_{t1-t2}}_{i}$ be its monthly incremental h-index.

The h-indices evaluate the influence of scientists by considering citations. In our case, h-indices enable a precise assessment of the significance of each author in establishing collaboration between two technologies. In fact, if an author has a high h-index, his impact on the collaboration between two technologies is assumed to be significant.

Let A^{A}_{t1-t2} be the average attribution score to $t1$ and to $t2$ of all the papers related to $t1$ or $t2$ published during a specific month having as an author. This quantifies the proximity of papers, for which A_{t1-t2} serves A_{t1-t2} as an author, to the technologies $t1$ and $t2$ within a designated month.

Let N^{A}_{t1-t2} be the average of the times A_{t1-t2} occurs in papers related to $t1$ and $t2$ published during a specific month. This variable significantly determines how relevant a given author's collaboration is to connect two technologies.

For a given month m, we define the two indices of proximity based on the collaboration of authors between two technologies, as follows:

$$IA^{m}_{t1-t2n} := \sum_{A_{t1-t2}} N^{A}_{t1-t2} * H^{A_{t1-t2}}_{n} * A^{A}_{t1-t2}$$

$$IA^{m}_{t1-t2i} := \sum_{A_{t1-t2}} N^{A}_{t1-t2} * H^{A_{t1-t2}}_{i} * A^{A}_{t1-t2}$$

These indices aim to measure how significant collaboration is for the proximity between technology $t1$ and $t2$. The approach we utilize involves the aggregation of coefficients assigned to each author, quantifying the significance and influence of their work within the connection between two technologies. This aggregation spans across all authors who have contributed to publications related to both technologies under consideration. As a result, this index evaluates the proximity of technologies based on collaboration.

Interpolation and Fitting Data

Once we've calculated all the proximity indices, the next step involves interpolating the time series of these indices. It was necessary because some technologies lack associated papers in certain months. Across the 625 time series, we achieve an average interpolation rate of around 20%, a reasonable percentage given that some technologies have only a limited number of attributed papers.

For the interpolation process, we employ a polynomial method of degree 3, which is a mathematical technique widely used in data analysis for approximating a curve or function. In practical terms, polynomial interpolation utilizes a polynomial function of a specified degree to approximate a function passing through a set of discrete data points. For example, a cubic polynomial is employed for interpolation of degree 3. Given four distinct data points (x_0, y_0), (x_1, y_1), (x_2, y_2), and (x_3, y_3), the formula for estimating the value y at some intermediate value of x, denoted as x_i, using cubic polynomial interpolation is expressed as follows:

$$y = a_0 + a_1 * (x_i - x_0) + a_2 * (x_i - x_0)^2 + a_3 * (x_i - x_0)^3$$

To determine the coefficients a_0, a_1, a_2, and a_3, a system of equations based on the given data points must be solved. Various methods, such as Lagrange interpolation or the Newton divided difference method, can be employed to derive these coefficients.

The simplest form of interpolation is linear interpolation, directly estimating values between two adjacent data points. This assumes a linear relationship between the data points. Specifically, for two points (x_1, y_1) and (x_2, y_2), the linear interpolation formula for estimating the value y at some intermediate value of x, denoted as x_i, is given by:

$$y = y_1 + (x_i - x_1) * \frac{(y_2 - y_1)}{(x_2 - x_1)}$$

In practice, one can vary the degree of interpolation based on the data's nature and the desired accuracy. Linear interpolation (degree one) is quick and simple but assumes a linear relationship, while higher-degree polynomial interpolation provides a more exact fit through data points but may introduce unnecessary oscillations. In our case, we initially chose a polynomial interpolation of degree 3 for the time series, offering flexibility without excessive oscillations. However, this leaves missing values at the time series' extremities, where some series start or end with gaps. To address this, we apply a second interpolation using a linear method.

This two-step process ensures all-time series are filled throughout the entire period. Negative values created by the interpolation are replaced with zeroes, as defined by the indices, which can only yield non-negative values.

Upon plotting some of the time series, we observe a cloud of points moving in a certain direction rather than forming a proper line, as depicted in Fig. 1.

The observed fluctuations in the time series can be attributed to various factors influencing our computations. These factors include computed variables such as cosine similarities or h-indices, as well as variables present in the OpenAlex dataset, like the attribution score to technologies based on the classification algorithm. Moreover, the dynamic nature of the scientific landscape, subject to monthly variations, contributes significantly to the time series' fluctuations. This dynamism arises from various causes, including varying publication frequencies for different technologies and the sudden integration of new scientific discoveries or algorithms leading to substantial citations by a significant portion of the scientific community, among other influences. Numerous factors contribute to why the computed time series are not entirely smooth. Nevertheless, the crucial aspect of our work lies in the general tendencies reflected by these indices.

To address the variability, we opt to fit curves to the points obtained for each time series. We perform polynomial interpolation, computing eleven polynomials for each time series with degrees ranging from 0 to 10. To determine the best fit for

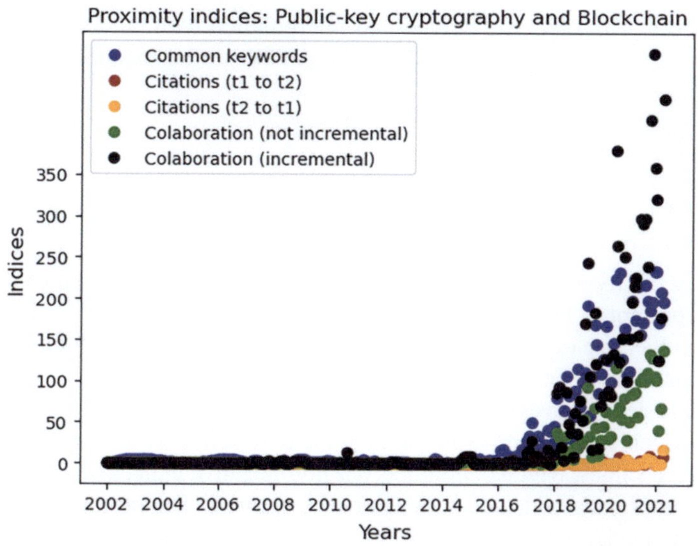

Fig. 1 Indices of proximity between Public-key cryptography and Blockchain from 2002 to 2021

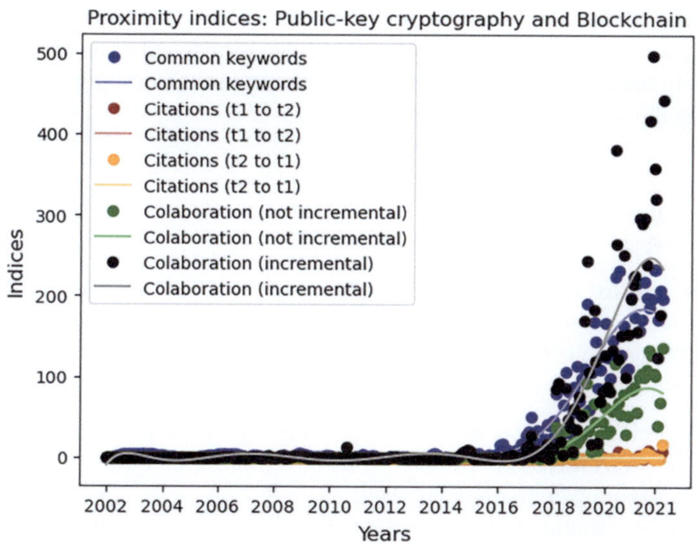

Fig. 2 Optimal polynomial fitting of the time series of proximity indices between Public-key cryptography and Blockchain with an interpolation rate of 24% from 2002 to 2021

the time series, we select the polynomial interpolation with the lowest Symmetric Mean Absolute Percentage Error (SMAPE). An illustrative example is provided in Fig. 2.

Clustering

In this study, our focus is on implementing clustering methods to rearrange time series data that represent technological proximity based on their shapes across various index types. The objective of this analysis is to pinpoint time series exhibiting increasing proximity, indicating potential technological convergence.

Our initial step involves preprocessing the time series. Following numerical experiments, we decide against de-seasonalizing the time series, as this introduces significant negative values, which contradicts our non-negative indices' definition. Subsequently, we normalize all-time series between 0 and 1. The normalization process begins with standardizing the time series data by subtracting the minimum value from each data point and then scaling the entire series by dividing it by the maximum value within the adjusted time series. This results in a time series that proportionally expresses the position between the minimum and maximum values of the original time series. Lastly, due to the high volatility of the normalized time series, we opt to smooth them using exponential smoothing with a robust smoothing parameter of $\alpha = 0.1$ using the following formula:

$$F_t = \alpha x_t + (1 - \alpha)F_{t-1},$$

where

- F_t is the forecast for the current time period t.
- x_t is the actual observation in the current period.
- F_{t-1} is the previous forecast.
- α is the smoothing parameter ($0 < \alpha < 1$).

The smoothing parameter α controls the weight given to the current observation versus the previous forecast. A smaller α assigns more weight to older data, while a larger α gives more weight to recent data.

Following the application of exponential smoothing, we exclude all-time series from our analyses with an interpolation rate exceeding 50%. These time series, accounting for 25% of the dataset, are considered artificial and represent noise in our dataset, potentially biasing our results.

Next, we allocate all normalized time series with a mean lower or equal to 0.02 to an artificial cluster representing a flat time series. This artificial cluster encompasses approximately 15% of the entire time series set. This approach allows the clustering algorithm to focus exclusively on data that cannot be manually clustered, constituting around 60% of our original time series.

We employ three clustering algorithms for time series: K-Means, K-Shape, and K-Medoids. We decided to fit these clustering algorithms in two ways: first, by training them in one part of our time series, and second, by training them on a large set of approximately 20,000 time series that we import from the Darts library.[2] To measure the distance between time series, we use the metric "l1", the sum of the absolute value of the differences between vectors or time series, as in our case. We do not want to penalize time series with large numerical differences too much, as this would be the case with the Euclidean distance, where the difference between points is squared, since the main goal is to cluster them by shape and not directly by the values they take.

To evaluate the performance of the clustering algorithm, we implement several strategies. We first use the silhouette score (Rousseeuw, 1987) to compute the clustering quality for each sample, each cluster, and finally for all the data. Then, we visualize the number of time series by cluster to see how the time series are distributed among the clusters. Afterward, we visualize the centroids of the clusters on a surface, with a positioning showing distances proportional to their distances as time series and with sizes proportional to the number of samples contained in each cluster. Finally, for each cluster, we plot randomly chosen time series within the cluster to assess their similarities visually.

To calculate silhouette score, one first calculates the average distance a_i from x_i for a given data point x_i to all other data points within the same cluster. This measures the cohesion of the data point with its cluster.

Then, one calculates the average distance b_i from x_i to all data points in the nearest neighboring cluster (i.e., the cluster other than the one to which x_i belongs). This measures the separation of the data point from other clusters.

The silhouette score s_i for the data point x_i is then calculated using the following formula:

$$\frac{s_i = (b_i - a_i)}{\max(a_i, b_i)}$$

[2] https://unit8co.github.io/darts/.

If the value of s_i is close to 1, it indicates that the data point is well-clustered and is far away from neighboring clusters. If the value of s_i is close to -1, it suggests that the data point is misclassified into the wrong cluster, as its distance to its own cluster is much greater than the distance to the nearest neighboring cluster. If the value of s_i is around 0, it means that the data point is on or very close to the boundary between two clusters. Regarding the overall silhouette score for an entire clustering solution (i.e., a set of clusters), one obtains it by computing the silhouette score for each data point in the dataset and then taking the average of these individual silhouette scores.

Ultimately, K-Shape with 5 clusters emerged as the optimal choice for our time series, achieving an average silhouette score of 0. This result indicates that the clusters are reasonably well-balanced, and the centroids exhibit sufficient separation from each other.

Forecasting

Our goal is to predict proximity indices and, ultimately, forecast how the closeness between encryption technologies evolves to predict technological convergence. We employ four methods to forecast time series data for 3, 6, and 12 months, aiming for optimal results.

In the first method, known as "local forecasting," each algorithm is trained on a segment of each time series, and then the given time series is directly forecasted with the fitted algorithm.

In the subsequent methods, "clustering forecasting" and "global forecasting," we categorize time series based on common keywords, citations, and collaborations. In clustering forecasting, each algorithm is trained on all-time series within a cluster, testing if using an algorithm trained on similar time series data leads to better forecasting. In global forecasting, each algorithm is trained on an all-time series, forecasting an all-time series with the fitted algorithm.

The last method, "transfer learning forecasting," involves training each algorithm on a large set of time series imported from Darts and then forecasting all-time series with the fitted algorithm.

Before forecasting, we preprocess the time series by cleaning the data to remove noise. We select time series with less than 50% interpolation, smooth them using exponential smoothing with a smoothing parameter of $\alpha = 0.1$, and start forecasts from December 2021, considering incomplete updates for 2022 in OpenAlex data.

To understand forecasting quality, we use the Symmetric Mean Absolute Percentage Error (SMAPE).

Let $A := \{A_t\}_{t\geq 1}^{n}$ and $F := \{F_t\}_{t\geq 1}^{n}$ be respectively the actual and the forecasted time series taking values at n time periods. The symmetric mean absolute percentage error between A and F is defined as follows:

$$\text{SMAPE}(A, F) := \frac{100}{n}\sum_{t=1}\frac{|F_t - A_t|}{|F_t| + |A_t|\frac{1}{2}}$$

If the actual and the forecasted time series are both zero at some time t, the contribution to the global sum is defined as zero. Using the triangle inequality, we obtain the following bound for SMAPE:

$$\text{SMAPE(A, F)} \leq 200$$

since

$$\begin{aligned}\text{SMAPE(A, F)} &\leq \frac{100}{n}\sum_{t=1}\frac{|F_t| + |A_t|}{(|F_t| + |A_t|)\frac{1}{2}}\\ &= \frac{100}{n}\sum_{t=1}\frac{1}{\frac{1}{2}}\\ &= \frac{100}{n}\left(\frac{n}{\frac{1}{2}}\right) = 200\end{aligned}$$

It is noteworthy that if the actual time series consistently remains zero while the forecasted time series consistently stays non-zero, even if the predicted time series closely aligns with the actual one, SMAPE between both time series will be 200%. Given the presence of numerous flat time series in our data, this observation should be kept in mind during our analysis of the results.

We employ k-fold cross-validation with an expanding window to assess forecasting quality. In this approach, each time-series is divided into successive sections. Training windows are created, starting from the first section and expanding at each step with the next section. Algorithms are trained on each training window, with forecasts initiated from the end of each window. The SMAPE between the forecasts and actual values is computed for each iteration, and the average of these SMAPEs represents the global error for a specific algorithm. This method is employed to mitigate overfitting.

To visualize the distribution of forecasting errors across all expanding windows, we generate a histogram illustrating the number of forecasts by error size in a single plot with distinct colors, as depicted in Fig. 3.

More concretely, we split the data into several sets. We take 80% of the data to train our algorithm, and 20% is left for testing the algorithm. Then, out of our training set, we take 80% of the data to tune the hyperparameters of our models and 20% to validate the results. We optimize the hyperparameters of each algorithm for each type of forecasting. Then, we train the models and forecast the time series using the k-fold cross-validation with an expanding window, as explained above. Last, to compare the different performances between the algorithms for each forecasting horizon and type of forecasting, we visualize the trade-off between the error and the computation time. This allows us to choose the optimal method for each specific forecasting task, as discussed in Sect. 5.1.

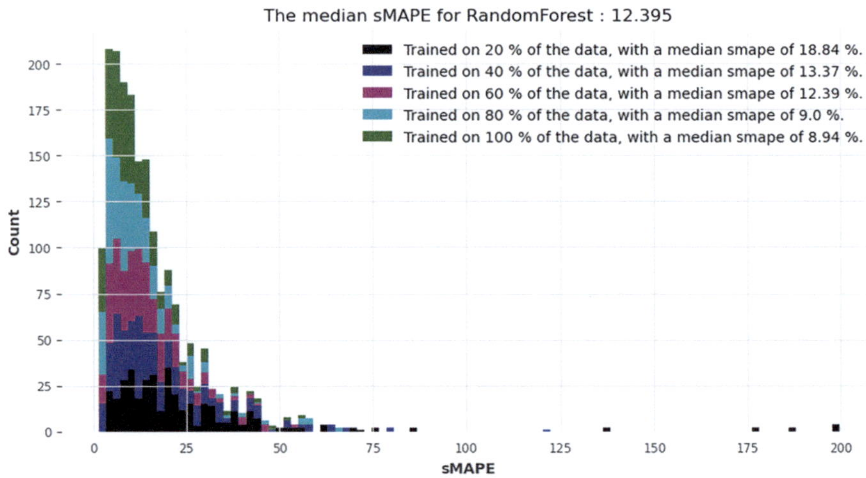

Fig. 3 Distribution of forecasting errors obtained by predicting time series of keyword-based proximity indices. The forecasting algorithm employed is Random Forest, trained globally on all our time series, with a forecasting horizon of 6 months

6 Results and Discussion

This section delves into the details of the results and their insights, followed by a case study illustrating the effectiveness of our proposed method in two cybersecurity topics. Finally, we discuss the limitations and future work.

Overall Results

The following tables provide detailed forecasting accuracy for each method, presenting data specific to index types, algorithms, and forecasting horizons. In these tables, Index 1 and Index 2 represent citations from one technology to another and vice versa. Indexes 3 and 4 denote collaboration based on incremental and non-incremental h-indices, respectively, while Index 5 pertains to common keywords. The colors brown, violet, and blue indicate the best forecasting accuracy for horizons of **3**, **6**, and **12** months, respectively (Table 1).

Overall, no single forecasting method consistently outperformed others across all tasks. For local forecasting, exponential smoothing and Theta algorithms demonstrated superior performance, while random forest outperformed linear regression and light gradient boosting machine (LGBM) in other forecasting types. In contrast, transfer learning methods exhibited comparatively poorer performance, potentially due to their computationally intensive nature and associated training limitations.

Different index types exhibited diverse forecasting outcomes. Citation indices, often flat, were straightforward to predict, resulting in a median SMAPEs of 0 across all forecasting horizons (see Table 4). Conversely, collaboration indices posed greater challenges, with the most accurate predictions stemming from randomized clustering forecasting using random forests (refer to Table 2). Keyword indices saw reasonable forecasting, with algorithmic clustering via random forests providing the most accurate results (Tables 3 and 5).

Table 1 Median SMAPEs for local forecasting

Algorithm	Horizon	Index 1	Index 2	Index 3	Index 4	Index 5
Naive seasonal (K=1)	3 months	3.89	3.89	20.95	20.82	6.70
	6 months	5.37	5.37	35.09	33.54	11.29
	12 months	8.28	8.28	54.56	50.78	14.97
Naive seasonal (K=12)	3 months	3.02	3.02	88.89	80.78	28.38
	6 months	5.24	5.24	87.75	82.94	28.17
	12 months	9.66	9.66	86.11	81.75	26.63
Exponential	3 months	1.53	1.53	20.39	20.22	7.05
	6 months	3.59	3.59	33.56	31.64	11.36
	12 months	10.10	10.10	57.66	54.18	15.81
Theta	3 months	1.99	1.99	19.91	20.06	7.08
	6 months	4.47	4.47	32.98	30.96	11.07
	12 months	8.18	8.18	54.97	50.69	14.50

Table 2 Median SMAPEs for randomized clustering forecasting

Algorithm	Horizon	Index 1	Index 2	Index 3	Index 4	Index 5
Linear Regression	3 months	74.09	33.65	116.38	128.49	28.36
	6 months	37.84	45.83	76.63	62.15	21.22
	12 months	32.31	26.54	78.06	69.87	17.95
LGBM	3 months	36.73	32.32	73.20	79.81	11.99
	6 months	38.91	30.48	48.72	51.47	11.61
	12 months	24.94	20.94	53.04	44.72	13.44
Random Forest	3 months	3.06	2.36	19.41	19.01	6.55
	6 months	4.19	2.46	28.80	27.10	9.55
	12 months	2.97	5.55	43.02	37.12	11.97

Table 3 Median SMAPEs for non-randomized clustering forecasting

Algorithm	Horizon	Index 1	Index 2	Index 3	Index 4	Index 5
Linear Regression	3 months	41.31	39.45	84.32	98.18	17.97
	6 months	75.34	35.25	82.19	82.75	16.42
	12 months	59.14	45.08	100.35	89.80	19.96
LGBM	3 months	59.17	15.89	80.05	108.49	17.44
	6 months	49.16	15.47	70.92	84.61	18.46
	12 months	50.32	21.86	56.16	85.09	14.36
Random Forest	3 months	8.24	4.47	41.61	31.41	8.26
	6 months	8.26	6.05	43.24	33.06	10.47
	12 months	11.94	5.11	65.71	38.29	12.08

Table 4 Median SMAPEs for global forecasting

Algorithm	Horizon	Index 1	Index 2	Index 3	Index 4	Index 5
Linear Regression	3 months	0.0	0.0	62.04	51.85	10.87
	6 months	0.0	0.0	70.17	63.72	14.08
	12 months	0.0	0.0	86.31	90.55	17.10
LGBM	3 months	0.0	0.0	31.37	33.17	10.22
	6 months	0.0	0.0	47.61	45.59	13.19
	12 months	0.0	0.0	59.60	52.98	16.09
Random Forest	3 months	0.0	0.0	22.01	21.62	8.25
	6 months	0.0	0.0	39.27	35.07	11.63
	12 months	0.0	0.0	54.55	49.82	16.08

Table 5 Median SMAPEs for transfer learning forecasting

Algorithm	Horizon	Index 1	Index 2	Index 3	Index 4	Index 5
Linear Regression	3 months	91.06	90.19	107.99	128.38	82.52
	6 months	92.06	91.44	118.68	135.36	102.58
	12 months	94.80	94.32	128.14	144.67	117.81
LGBM	3 months	98.47	97.93	165.79	171.97	182.81
	6 months	98.50	97.93	166.26	172.05	183.46
	12 months	98.62	98.08	167.35	172.94	184.71
Random Forest	3 months	98.18	97.62	164.16	170.19	182.06
	6 months	98.03	97.70	161.82	170.404	178.98
	12 months	98.31	97.67	160.16	169.59	180.76

Case Study

To illustrate the practical implications of our findings, we examine the evolution of proximity indices between "Public-key cryptography" and "Blockchain," as depicted in Fig. 4.

A notable interpolation rate of 46% is evident. This can be attributed to the sparse interactions between both technologies from 2002 to 2012, reflecting the early stages of blockchain development in the cybersecurity domain. Consequently, during this phase, indices remained relatively stagnant. It's important to note that this interpolation rate, though steep, doesn't significantly bias our indices between "Public-key cryptography" and "Blockchain"; instead, it accurately reflects the non-interactive phase between these technologies.

A significant surge in the proximity indices is evident from 2017 onwards, with the indices for common keywords and collaboration based on incremental h-indices being particularly prominent. This suggests a robust correlation between the variables "common keywords" and "collaboration," indicating a notable number of authors simultaneously exploring both public-key cryptography and blockchain technologies. Additionally, there is a marginal increase in mutual citations between the two technologies. While citation-based indices exhibit modest growth, it could be attributed to their inherent construction rather than a lack of interaction between the two technologies. Based on this evidence, we hypothesize that a tangible technological convergence occurred between blockchain and public-key cryptography from 2017 to 2021.

This convergence is highlighted by the increasing adoption of public-key cryptography in blockchain platforms during this period. Techniques like digital signatures, based on public-key cryptography, became integral to verifying blockchain transactions. Furthermore, the integration of zero-knowledge proofs, also based on public-key cryptography, gained traction in blockchain frameworks to validate assertions without revealing the underlying data. In conclusion, our findings highlight a pivotal phase of technological convergence between blockchain and public-key cryptography spanning 2017–2021.

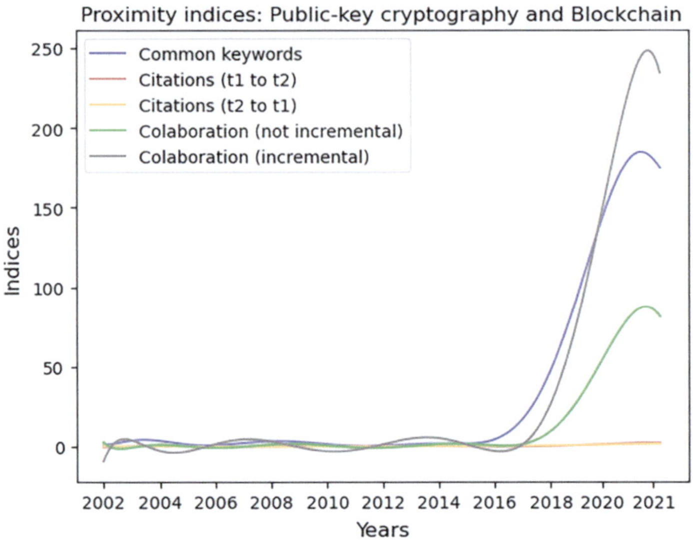

Fig. 4 Case study: Examining the technological proximity between public-key cryptography and blockchain. The proximity indices undergo interpolation and are modeled using polynomial curves

Limitations and Future Works

While our research offers valuable insights, it also has limitations and suggests areas for future work. One limitation of our proposed method is its reliance on a single data source, OpenAlex, without integrating or comparing it to other similar datasets. This could potentially result in our proposed method being overfitted toward OpenAlex. Additionally, our approach lacks a normalization strategy for various calculations of proximity indices. As a result, index comparisons may be influenced by predominant values, such as the high number of publications in a specific technology domain. To address this issue, we propose index normalization by averaging monthly publications relevant to the technologies or using a weighted factor based on available variables.

Introducing a normalized index as an additional graph edge could enhance the computation of different proximity indices. This additional edge, forming five-dimensional edges instead of our quintet of distinct indices, could contribute to establishing a more improved convergence threshold. Subsequently, leveraging community detection algorithms might provide insights into technological convergence by identifying clusters of converging technologies. This avenue holds promises for further exploration.

7 Conclusion

In this study, we introduce a method that uses text mining and bibliometrics techniques, utilizing the "OpenAlex" catalog, to create and predict technological proximity indices specific to encryption technologies. A case study applying our method highlights the convergence between blockchain and public-key cryptography. The insights gained from our method can offer valuable guidance for stakeholders and enthusiasts navigating the impacts of encryption technology.

We recognize certain constraints in our study. Our literature corpus was limited to "OpenAlex," which may not comprehensively represent the core interests and activities in the research community working on encryption technologies. Additionally, the non-normalized nature of our indices hinders direct comparisons among various data sources. These limitations also suggest potential paths for future research to expand our findings, enriching the field of scientific monitoring within the encryption technology community.

Supplementary information The code of this project is on https://github.com/technometrics-lab/Proximity-indices-applied-to-OpenAlex.

Acknowledgements We thank Jacques Roitel, whose work was the original base of this paper. Additionally, we would like to thank the other interns working at armasuisse at the time this research was led for their helpful coding advice: Léo Meynent, Marc Egli, Francesco Intoci, Lucas Crijns, and Alexander Glavackij.

Appendix

This appendix includes figures, graphics, and bar charts that were referenced but not directly included in the main paper. They are presented in the order of their mention in the paper (Figs. 5, 6, 7, 8, 9 and 10).

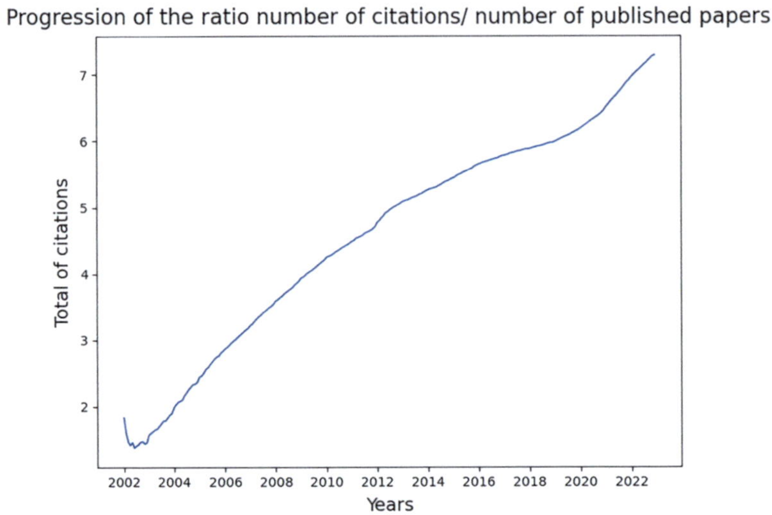

Fig. 5 Evolution of the ratio between the number of citations and the number of published papers in the field of encryption technologies

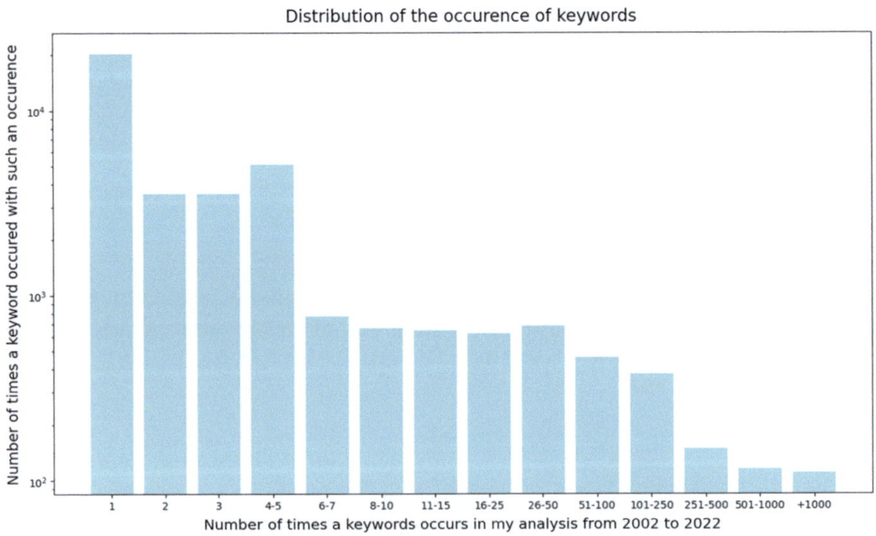

Fig. 6 Distribution of the occurrence of keywords among all papers related to encryption technologies published between 2002 and 2022

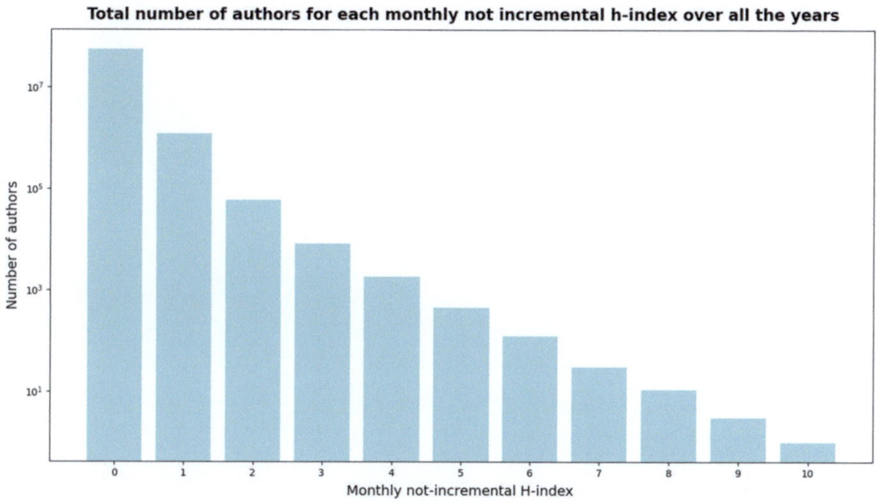

Fig. 7 Distribution of the monthly non-incremental h-indices of all the authors of papers related to encryption technologies between 2002 and 2022

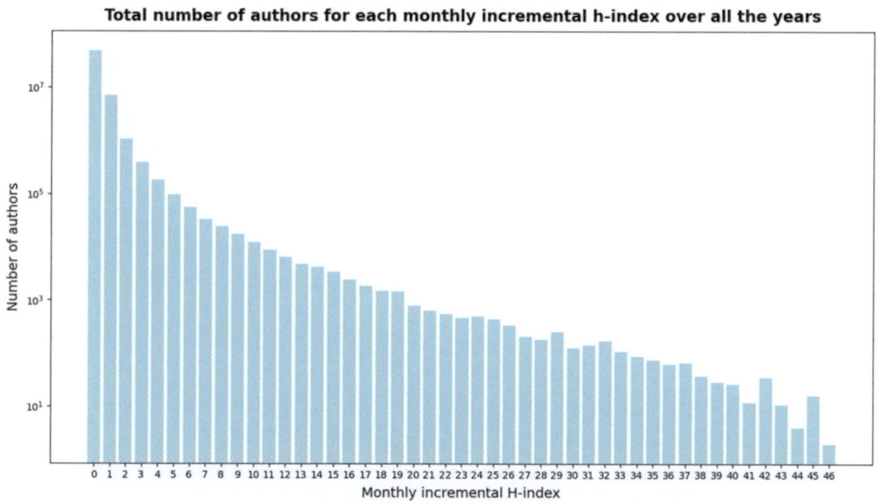

Fig. 8 Distribution of the monthly incremental h-indices of all the authors of papers related to encryption technologies between 2002 and 2022

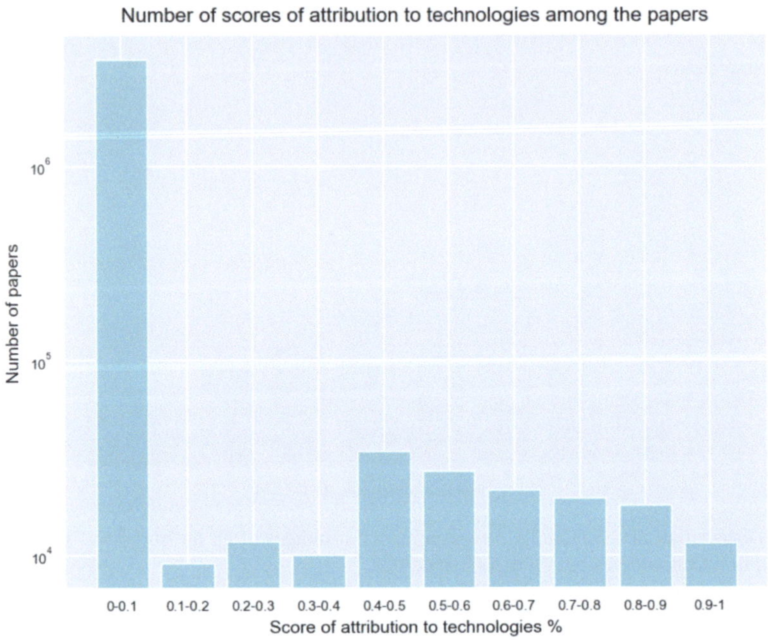

Fig. 9 Distribution of the score of attribution to concepts given by OpenAlex among all papers related to encryption technologies published between 2002 and 2022

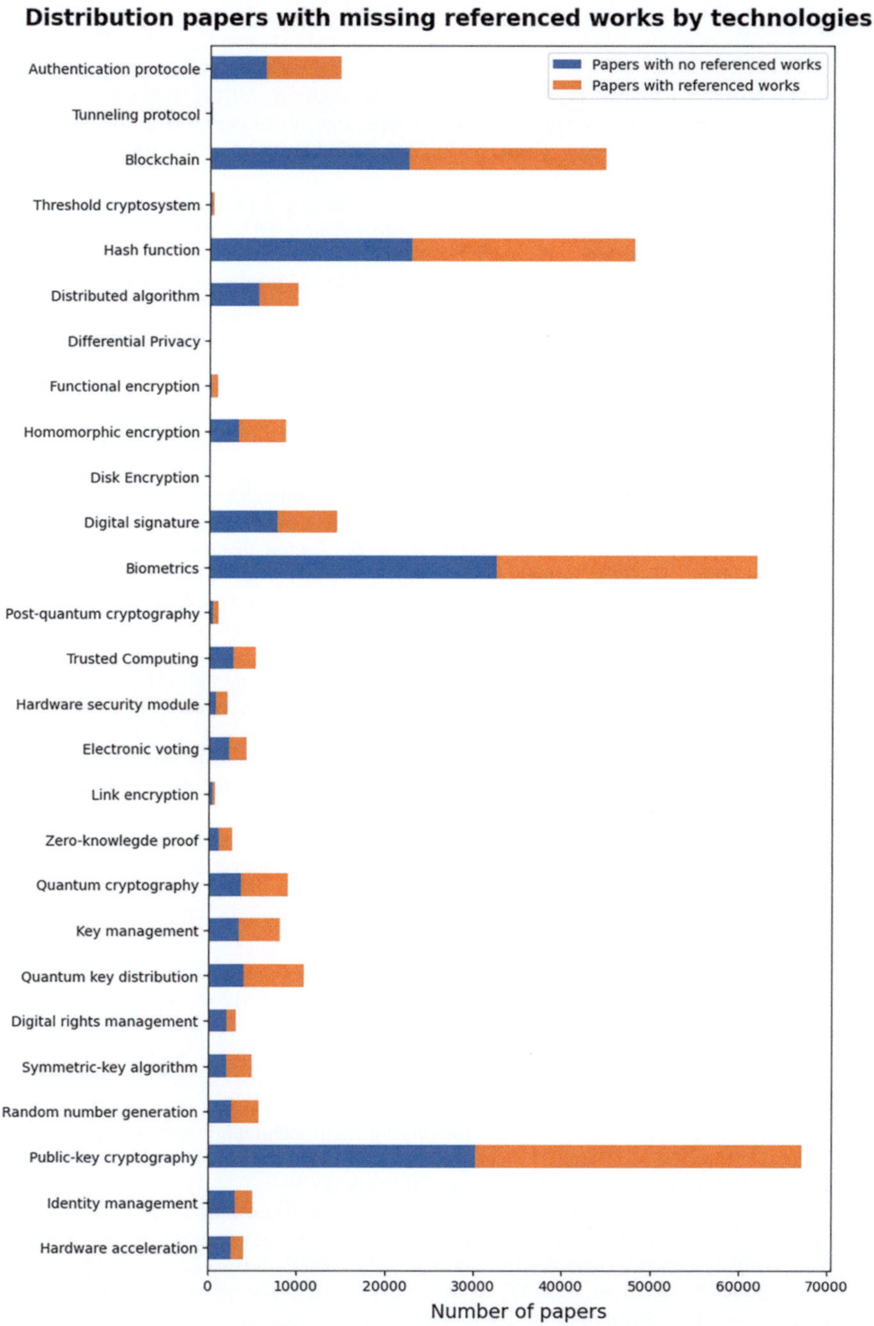

Fig. 10 Distribution of the papers with missing referenced works by technologies for all the papers related to encryption technologies published between 2002 and 2022

References

Blouin, M., Sery, N., Cluzeau, D., Brun, J.-J., & Bédécarrats, A. (2013). Balkanized research in ecological engineering revealed by a Bibliometric analysis of earthworms and ecosystem services. *Environmental Management, 52*(2), 309–320. https://doi.org/10.1007/s00267-013-0079-8.Accessed2022-10-19

Breitzman, A., & Thomas, P. (2015). The emerging clusters model: A tool for identifying emerging technologies across multiple patent systems. *Research Policy, 44*(1), 195–205. https://doi.org/10.1016/j.respol.2014.06.006.Accessed2022-10-21

Brown, B. B., et al. (1968). *Delphi process: A methodology used for the elicitation of opinions of experts*. Rand Corporation Santa Monica.

Carley, S. F., Newman, N. C., Porter, A. L., & Garner, J. G. (2018). An indicator of technical emergence. *Scientometrics, 115*(1), 35–49. https://doi.org/10.1007/s11192-018-2654-5.Accessed2022-09-30

Chand Bhatt, P., Kumar, V., Lu, T.-C., & Daim, T. (2021). Technology convergence assessment: Case of blockchain within the IR 4.0 platform. *Technology in Society 67*, 101709. Accessed 20 Oct 2022. https://doi.org/10.1016/j.techsoc.2021

Chen, H., Zhang, Y., & Zhu, D. (2016). Identifying technological topic changes in patent claims using topic modeling. In T.U. Daim, D. Chiavetta, A.L. Porter, & O. Saritas (Eds.) *Anticipating Future Innovation Pathways Through Large Data Analysis. Innovation, Technology, and Knowledge Management* (pp. 187–209). Springer, Cham. https://doi.org/10.1007/978-3-319-39056-7 11

Cojocaru, S., Bragaru, C., & Ciuchi, O. (2012). The role of language in constructing social realities. The Appreciative Inquiry and the reconstruction of organizational ideology. *Revista de Cercetare si Interventie Sociala 36*, 31–43.

Curran, C.-S., Br¨oring, S., & Leker, J. (2010). Anticipating converging industries using publicly available data. *Technological Forecasting and Social Change* **77**(3), 385–395. Accessed 21 Oct 2022. https://doi.org/10.1016/j.techfore.2009.10.002

Halaweh, M. (2013). Emerging technology. *What is it Journal of Technology Management &Amp; Innovation 8*(3), 108–115. Accessed 12 Sep 2022. https://doi.org/10.4067/S0718-27242013000400010

He, C., Shi, F., & Tan, R. (2022). A synthetical analysis method of measuring technology convergence. *Expert Systems with Applications, 209*, Article 118262. https://doi.org/10.1016/j.eswa.2022.118262.Accessed2022-10-21

Huang, L., Cao, X., Cai, Y., Ren, H., Xing, T., & Ye, P. Detecting technological recombination using semantic analysis and dynamic network analysis 11.

Kajikawa, Y., & Takeda, Y. (2008). Structure of research on biomass and bio-fuels: A citation-based approach. *Technological Forecasting and Social Change, 75*(9), 1349–1359. https://doi.org/10.1016/j.techfore.2008.04.007.Accessed2022-10-18

Kim, J., Kim, S., & Lee, C. (2019). Anticipating technological convergence: Link prediction using Wikipedia hyperlinks. *Technovation, 79*, 25–34. https://doi.org/10.1016/j.technovation.2018.06.008.Accessed2022-10-21

Kim, J., & Lee, S. (2017). Forecasting and identifying multi-technology convergence based on patent data: The case of IT and BT industries in 2020. *Scientometrics, 111*(1), 47–65. https://doi.org/10.1007/s11192-017-2275-4.Accessed2022-10-21

Kim, T. S., & Sohn, S. Y. (2020). Machine-learning-based deep semantic analysis approach for forecasting new technology convergence. *Technological Fore- Casting and Social Change, 157*, Article 120095. https://doi.org/10.1016/j.techsoc.2020.120095.Accessed2022-10-20

Klarin, A., Suseno, Y., & Lajom, J. (2021). Systematic literature review of convergence: A systems perspective and re-evaluation of the convergence process. *IEEE Transactions on Engineering Management*, 1–13. https://doi.org/10.1109/TEM.2021.3126055

Kumar, T., Vaidyanathan, S., Ananthapadmanabhan, H., Parthasarathy, S., & Ravindran, B. (2020). Hypergraph clustering by iteratively reweighted modularity maximization. *Applied Network Science, 5*(1), 1–22. https://doi.org/10.1007/s41109-020-00300-3.Number:1Publisher:SpringerOpen.Accessed2022-10-20

Lee, C., Kwon, O., Kim, M., & Kwon, D. (2018). Early identification of emerging technologies: A machine learning approach using multiple patent indicators. *Technological Forecasting and Social Change, 127*, 291–303. https://doi.org/10.1016/j.techfore.2017.10.002.Accessed2022-10-17

Li, H., An, H., Wang, Y., Huang, J., & Gao, X. (2016). Evolutionary features of academic articles co-keyword network and keywords co-occurrence network: Based on two-mode affiliation network. *Physica a: Statistical Mechanics and Its Applications, 450*, 657–669. https://doi.org/10.1016/j.physa.2016.01.017.Accessed2022-09-09

Liu, P., & Xia, H. (2015). Structure and evolution of co-authorship network in an interdisciplinary research field. *Scientometrics, 103*(1), 101–134. https://doi.org/10.1007/s11192-014-1525-y.Accessed2022-11-30

Lu¨, L., & Zhou, T. (2011). Link prediction in complex networks: A survey. *Physica A: Statistical Mechanics and its Applications 390*(6), 1150–1170. Accessed 17 Oct 2022. https://doi.org/10.1016/j.physa.2010.11.027.

MacRoberts, M. H., & MacRoberts, B. R. (1996). Problems of citation analysis. *Scientometrics, 36*(3), 435–444. https://doi.org/10.1007/BF02129604.Accessed2022-11-23

Mitrea, C., Lee, C., & Wu, Z. (2009). A comparison between neural networks and traditional forecasting methods: A case study. *International Journal of Engineering Business Management 1*. https://doi.org/10.5772/6777

Porter, A. L., Garner, J., Carley, S. F., & Newman, N. C. (2019). Emergence scoring to identify frontier R&D topics and key players. *Technological Forecasting and Social Change, 146*, 628–643. https://doi.org/10.1016/j.techfore.2018.04.016.Accessed2022-09-13

Priem, J., Piwowar, H., & Orr, R. (2022). OpenAlex: A fully-open index of scholarly works, authors, venues, institutions, and concepts. Accessed 03 Oct 2022. arXiv:2205.01833. https://doi.org/10.48550/arXiv.2205.01833. http://arxiv.org/abs/2205.01833

Rousseeuw, P. J. (1987). Silhouettes: A graphical aid to the interpretation and validation of cluster analysis. *Journal of Computational and Applied Mathematics, 20*, 53–65.

Schoen, A., Villard, L., Laurens, P., Cointet, J.-P., Heimeriks, G., & Alkemade, F. (2012). The network structure of technological developments; technological distance as a walk on the technology map.

Song, C. H., Elvers, D., & Leker, J. (2017a). Anticipation of converging technology areas—A refined approach for the identification of attractive fields of innovation. *Technological Forecasting and Social Change, 116*, 98–115. https://doi.org/10.1016/j.techfore.2016.11.001.Accessed2022-09-30

Song, K., Kim, K. S., & Lee, S. (2017b). Discovering new technology opportunities based on patents: Text-mining and F-term analysis. *Technovation, 60–61*, 1–14. https://doi.org/10.1016/j.technovation.2017.03.001.Accessed2022-09-30

Wang, Z., Porter, A. L., Wang, X., & Carley, S. (2019). An approach to identify emergent topics of technological convergence: A case study for 3D printing. *Technological Forecasting and Social Change, 147*, 723–732. https://doi.org/10.1016/j.techfore.2018.12.015.Accessed2022-11-24

Zhou, Y., Dong, F., Kong, D., & Liu, Y. (2019). Unfolding the convergence process of scientific knowledge for the early identification of emerging technologies. *Technological Forecasting and Social Change, 144*, 205–220. https://doi.org/10.1016/j.techfore.2019.03.014.Accessed2022-09-30

Detection of Emerging Technologies

Shino Iwami

1 Introduction

This report adds standards and open source software (OSS) to research papers and patents as intellectual property and explains the significance of intellectual property in the economy. Easy methodologies for detecting emerging technologies and convenient methodologies for understanding technology overviews are introduced with the results about science and OSS. Additionally, this describes cases of strategies for capturing the market when you have a promising technology.

1.1 Types of Intellectual Properties

Many papers and patents have been analyzed in the fields of scientometrics and bibliometrics. In addition to being used as a survey before starting research or filing a patent application, they are now also being used to conduct evidence-based science and technology policy and innovation management (Leydesdorff & Milojević, 2012; Leydesdorff et al., 2014; Teich, 2018). Meanwhile, in the third industrial revolution of the twentieth century, though there was no unified definition, as the service industry developed, computers were introduced, automating factories and improving the efficiency of services. As automation and efficiency improved, the standards for machine parts were unified, and the number of interchangeable parts that could be used without depending on the manufacturer increased. For example, image and video formats such as MPEG (Moving Picture Experts Group) (Gall, 1991) have been patented but standardized as international standards so that users can commonly apply the standards to their products (Goedhuys & Sleuwaegen, 2013). Some of these

S. Iwami (✉)
NEC Corporation, Tokyo, Japan
e-mail: iwami.research@outlook.com

H. da Hora et al. (eds.), *Technology Mining*,
https://doi.org/10.1007/978-3-032-10849-4_4

Table 1 Scope by intellectual property

Intellectual Property	Scope			
	- Humanities and sociology - Theory in STEM	- Technology (Medical, agriculture, etc.) - Business model	Engineering	
			Hardware	Software
Research paper	○	○	○	○
Patent		○	○	○
Standard			○	○
OSS				○

standards are also a kind of intellectual property. While standards form a unified set of intellectual property for multiple organizations, Open source software (OSS) has emerged for intellectual property that can be shared with other organizations and individuals but does not require unified standards. OSS (Raymond, 2000) is the most successful intellectual property in the open culture. Additionally, open innovation (Chesbrough, 2003), open data, open governance (Chan, 2013), and other open trends are spreading. Table 1 shows the scope of four types of intellectual property: research paper, patent, standard, and OSS. Since the origin of modern universities started in the eleventh century (Tucker, 2023),[1] and peer review was conducted in the seventeenth century (Chapelle, 2014; Kronick, 1990), it is thought that the current process of research papers was in the seventeenth century. Regarding patents, in 1474, the first patent law was issued in Venetia (Fusco, 2019). The Institute of Electrical and Electronics Engineers (IEEE) proposed its first recommendation for the practical unit in 1890 (Engineering and Technology History Wiki, 2023), the International Electrotechnical Commission (IEC), one of the international standards, started in 1906 (IEC, 2023), and the International Organization for Standardization (ISO) was launched in 1947 (Tranchard, 2023). Thus, international standards were established approximately in the twentieth century. The concept of OSS was also developed in the twentieth century (Hippel & Krogh, 2003). At present, the OSS community is expanding its scope within software development to standards (OpenChain, 2023) and best practices (Open Source Security and Foundation, 2023),[2] such as activities about security related to the software supply chain.

As standards are considered to exist between OSS side and the side of research papers and patents, it was decided to focus on four intellectual properties, as shown in Table 1. Regarding intellectual property, trademarks, designs, and copyrights also exist. This report excludes non-technical intellectual property. However, since Apple and Sony are selling technological products with futuristic and stylish appearances and the concept of opening blueprints called open source hardware (Bonvoisin et al., 2017; Pearce, 2015) has emerged, it is kept in mind that designs may occupy the boundary bridging between technology and non-technology.

[1] The University of Bologna in Italy is famous as the first modern university, but higher education institutions different from modern ones existed outside of the West before the University of Bologna.

[2] IT was called "CII Best Practices badge" until 2021–12-24.

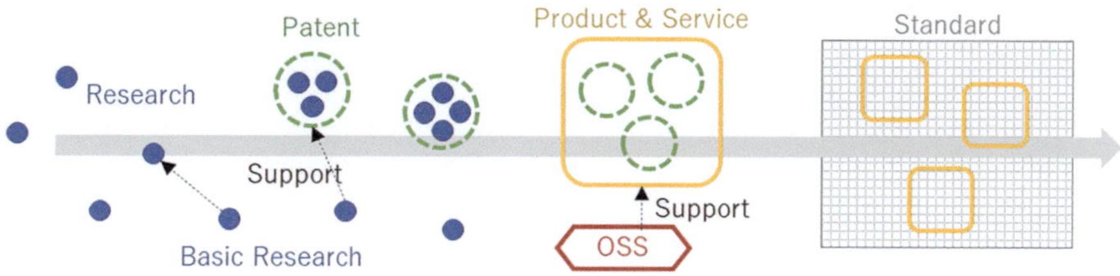

Fig. 1 Process of technical intellectual property from research to industrialization

Considering the process by which these four intellectual properties become products and services, it can be expressed in a different way from Table 1, as shown in Fig. 1. Several pieces of research may come together to become a patent, and the patents may be incorporated into a product or service. Standards are applied to products and services. Meanwhile, among research, some pieces of basic research become the foundation for everything, supporting other intellectual property, products, and services. Additionally, OSS may be incorporated into products and services, and may be the foundation for everything.

In the future, all data, such as audio and video, and artificial intelligence (AI) models, will be required to be properly managed as intellectual property or components for use in data science and AI. Currently, there is a global discussion on protecting such data as intellectual property and transparency through openness[3] (AI Alliance, 2024; The Linux Foundation, 2024a). This report does not cover such data, as the economic effects are still unknown and analysis in text format is difficult.

1.2 Significance of Intellectual Properties for the Economy

From the perspective of economic development, as shown by the Solow model (Solow, 1956), intellectual property is important as it enhances production efficiency. The Solow model says that capital stock, labor force, and productivity by high technologies are three important factors for economic growth. Profits from economic growth increase capital stock and are invested in improving labor and productivity. It is thought that these cycles promote economic growth as shown in Fig. 2. Solow won the Nobel Prize in Economics in 1987 (Nobel Prize Outreach, 1987). Intellectual properties are deeply related to efficiency by technological development, and the identification of emerging technologies is the key factor. In addition, intellectual property has a characteristic that provides benefits to first movers (Tarver, 2023). In order to realize a strong economy, there are ways to maintain possession of intellectual property as well as create and incorporate emerging technologies.

[3] For AI, Explainable AI (XAI) (Barredo Arrieta et al., 2020; Longo et al., 2024), Responsible AI, or other openness of AI are pursued in several AI ethics research.

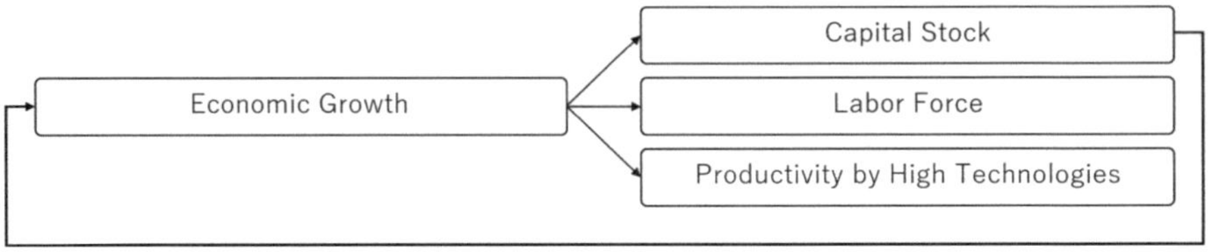

Fig. 2 Cycle of economic growth

Looking at growth accounting, for example, in the United States, total factor productivity (TFP) containing intellectual property has not contributed to economic growth since 2006. Some say this is because the software is not registered as intellectual property (Shimizu, 2019). It was difficult to calculate how to count the economic effects of OSS.[4] In this way, OSS is recognized as intellectual property, but it is confusing. Furthermore, it will be partially that the standard will be recognized as intellectual property. Meanwhile, it is not enough simply to have intellectual property and emerging technology. It is often argued that one factor of the economic decline of Japan was due to neglecting to create industries in the 1970s while the information technology (IT) industry, represented by Microsoft and Apple, was emerging in the United States of America. This claims the significance of utilizing new technologies for economic growth. As examples where technology alone is not enough, Japanese companies (Sony and Panasonic) had been ahead in music players and e-book readers, and then American companies (Apple and Amazon) provided comprehensive services that included not only hardware but also content. Products gained a larger share. Consequently, if the link between the creation of technologies and industrialization is weak, the creation of technologies ends up at a cost. The process from research and development that creates technology to industrialization is a serious issue, but it should be discussed separately in the future. Anyway, here the theme will concentrate on detecting emerging technology.

1.3 Demands for Detection of Emerging Technologies

Intellectual properties are important, but they are not valid for all time. Intellectual property will not work if two points do not match: the stage (newness) of the intellectual property and the acceptability of society (window of opportunity (Suarez et al., 2015)). This report focuses on the former, the newness, and technologies in the first stage are called emerging technologies. Entrants at earlier stages take a larger advantage, as mentioned in the first mover advantage (Lieberman & Montgomery, 1988), in exchange for higher risk. As time passed, emerging technologies are used more broadly, competitors also entered the market, and the profit will be down without

[4] From an industry-wide perspective, the value of OSS was estimated as 4.15 billion dollars on the supply side and 8.8 trillion dollars on the demand side in 2024 (Tarver 2023).

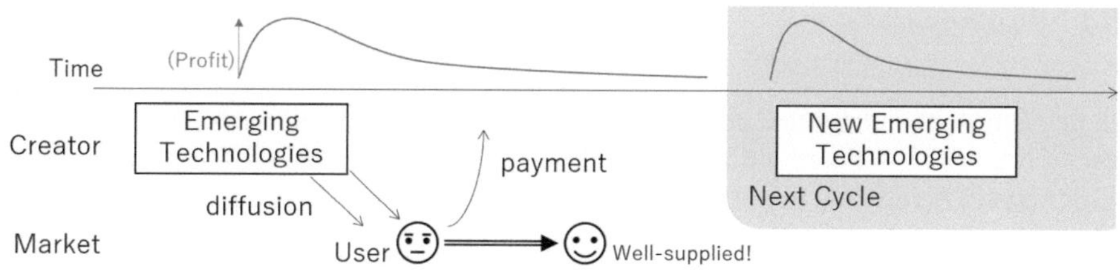

Fig. 3 Economic effects of emerging technologies

counterplans. The creators must release next emerging technologies for keeping profit by new emerging technology, as shown in Fig. 3. Therefore, organizations that value emerging technologies must maintain the detection of emerging technologies going to incubate new industries.

Several ways of thinking were devised for the lifetime of technology. The innovation theory proposed five stages (Rogers, 2003), and the Gartner Hype Cycle having five stages is famous (Gartner, 2024). The Cloud Native Computing Foundation (CNCF) manages projects in four stages (The Linux Foundation, 2024b). In business, the first stage may be intentionally avoided, as in the case of generic drugs, but the second stage and beyond cannot begin without the first stage, so it is important to detect emerging technologies in the first stage. A previous review paper referred to methodologies to identify emerging technologies (Rotolo et al., 2015), but they are limited to academic papers and patents. However, effects by other types of intellectual properties cannot be missed for economic growth. Thus, this report demonstrates available methodologies for the other types of intellectual properties as described in Sect. 1.2, and new keywords and new fields in the analyses are detected as emerging technologies as emerging technologies. Meanwhile, intellectual properties are not useful without society's acceptance. For example, electric vehicles were created before gasoline-powered cars were created (Ajanovic, 2015). Once electric vehicles disappeared, they appeared again as a solution to global environmental issues. It would be good to understand society's acceptance as well, but this report will not pursue it anymore to focus on emerging technologies.

Considering what emerging technologies are, most of them are derived from other technologies (Adner & Levinthal, 2002). However, a small number of things, such as green fluorescent protein (Bradford, 2024) and quasicrystals (Jeevan & Ranganathan, 2004), are discovered through chance events known as serendipity.[5] While it is difficult to observe the developmental process of emerging technologies from the inside, like organizations' and creators' thoughts, data makes it possible to identify them from the outside. The methodology demonstrated in this report is ubiquitous to observe from the outside.

[5] Serendipity is discoveries in science based on knowledge and observation (Copeland 2019; Gest 1997).

1.4 Purpose

In the following chapters, analysis methodologies and results are presented for research and OSS. Patents have the characteristics of (1) having citation relationships and (2) being able to extract keywords through natural language processing, so they can be analyzed in the same way as research. However, due to the characteristic of being moderately cited compared to papers, it is not possible to obtain as dense results between patented technologies as research. In addition, since Standard does not include citation-related information, it is thought that the same analysis as OSS is possible by extracting keywords using natural language.

From the point of view of analysis, patents, standards, and OSS are advantageous in that many databases are open to the public and it is easy to obtain data. On the other hand, OSS is advantageous in that it is easy to obtain the latest information because OSS includes a lot of information technologies whose rapid growth speed is named "dog year" (Wellman, 2001).

The rest of this report is organized as follows: Sect. 2 describes the methodologies. Section 4 provides results of quantitative analysis to detect emerging technologies and Sect. 5 discusses the results. Section 6 adds the strategies to after detecting emerging technologies, and this section provides additional information to use the results with the premise that our future can be led consciously as show in Fig. 4. Section 7 summarizes the report.

2 Methodologies

This section consists of two parts: the methodology for emerging keywords by scoring, and the methodology for overviewing by clustering networks. Faded keywords are computed as a supplement of emerging keywords. Figure 5 shows overviewing methodologies. In addition, technologies that evolve quickly, such as IT, are more appropriately classified as short-term emerging technologies, while technologies that evolve more slowly, such as materials, are more appropriately classified as middle-term emerging technologies. Overview methodologies using network clustering use big data, and therefore can obtain more reliable and stable results. However, when using citation networks, such as for papers and patents, it takes time for them to be cited in addition to the time it takes to publish. Networks using keyword relations,

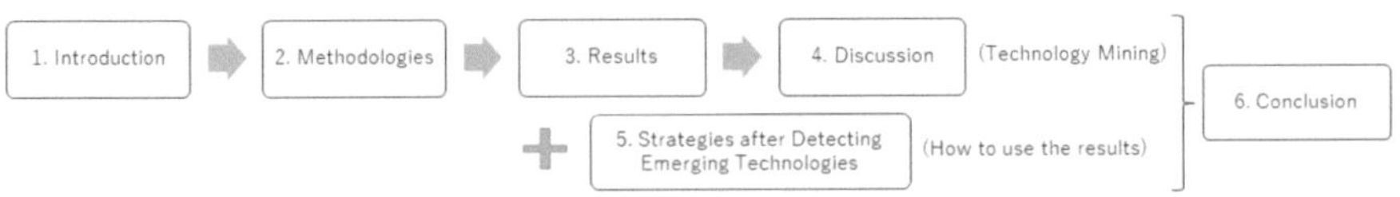

Fig. 4 Structure of the report

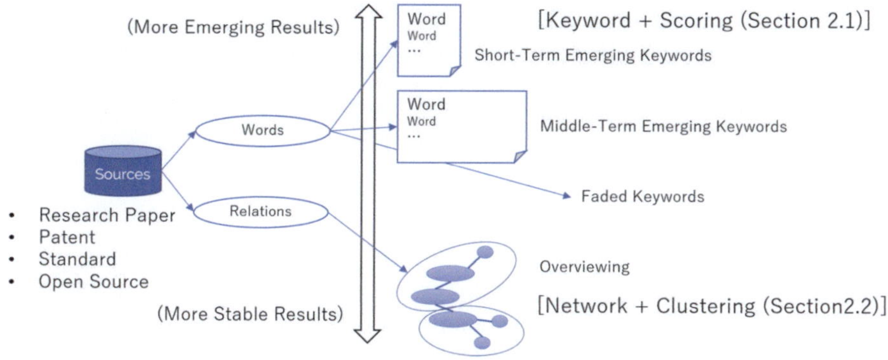

Fig. 5 Breakdown of methodologies

such as for OSS, can be analyzed immediately after publication, and results can be obtained more quickly than with the analysis with citations.

The advantage of using keywords in both scoring and clustering is that they can be used for any data. Analysis with citation networks is more difficult than that with texts because the data is more limited. Depending on the ideas, text analysis can also be used with non-intellectual property data, such as newspaper articles or social media. Of course, there are disadvantages, as discussed in Sect. 4.2 and after, but they do not affect the results in this report. For practical use, a few methodologies will give more reliable clues, though even one methodology can give clues to detect emerging technologies.

2.1 *Capturing Emerging Keywords*

There are methodologies for short-term changes and for middle-term changes. The analysis for the short-term changes is performed by the following steps:

(1) Keywords are retrieved from data sources.
(2) The keywords are counted by period.
(3) First appearance keywords are identified, and ranked by the appearance times.

The analysis for the middle-term changes is performed by the following steps:

(1) Keywords are retrieved from data sources.
(2) The keywords are counted by period.
(3) Growth rate by keyword is computed, and ranked by growth rate. The growth rate is defined as the count of the most recent period divided by the count of the previous period.

In addition, faded keywords are analyzed by the following steps:

(1) Keywords are retrieved from data sources.

(2) The keywords are counted by period.
(3) Disappearing keywords are identified, and ranked by the appearance times in the last period.

Steps (1) and (2) are executed at the same time.

2.1.1 Data Source

In the analyses of research papers, author keywords and KeyWords Plus on Clarivate Web of Science are available. The author keywords were set by the authors, and Clarivate added keywords as KeyWords Plus. The data of research papers are retrieved from the website in the format of tab-separated values (TSV). Keywords were limited to ones in the observation year, 2015.

In the analyses of OSS, topics on GitHub are used as keywords. The data of OSS are retrieved via GitHub API (GitHub, 2023), which provides topics for each OSS repository. The OSS repositories are limited to ones having 200 or more stars. However, no topic was set for some pieces of OSS. This analysis uses data from the in-organization system that runs periodically. Thus, the amount of data increases at the timings for updating.

Representative data sources for the four types of intellectual property are the following:

- Research Paper
  - Web of Science—https://www.webofknowledge.com/
  - Scopus—https://www.scopus.com/
  - Google Scholar—https://scholar.google.com/
- Patent
  - World Intellectual Property Organization (WIPO)—https://patentscope2.wipo.int/search/ja/search.jsf
- Standard
  - International Organization for Standardization (ISO)—https://www.iso.org/
  - International Electrotechnical Commission (IEC)—https://www.iec.ch/
  - International Telecommunication Union (ITU)—https://www.itu.int/
- OSS
  - GitHub— https://github.com/
  - SourceForge— https://sourceforge.net/

In the analyses of this report, data from the Web of Science and GitHub are mainly used.

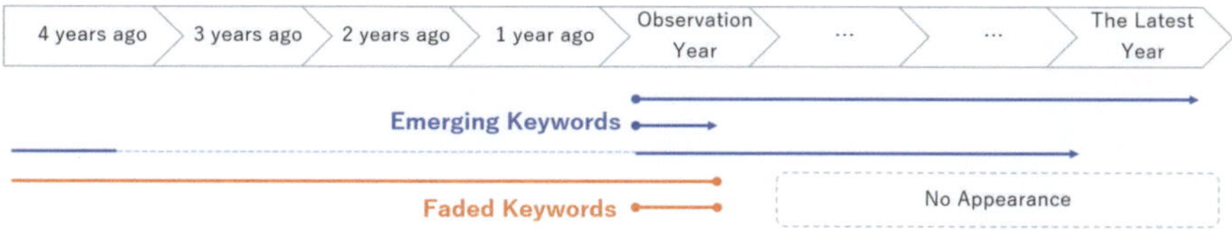

Fig. 6 Appearance periods of emerging and faded keywords

2.1.2 Definition of Short-Term Emerging Keywords and Faded Keywords

In the analyses of research papers, emerging keywords are defined as keywords that have appeared in the observation year. Faded keywords were not analyzed.

In the analyses of OSS, emerging keywords are defined as keywords that have not existed in the last three years. Thus, emerging keywords might appear four years ago. Meanwhile, faded keywords are defined as keywords that do not appear until the current year after the last year of appearance. Figure 6 shows patterns of timelines of emerging and faded keywords.

2.1.3 Computation of Middle-Term Emerging Keywords

Keywords having a high growing rate are defined as middle-term emerging keywords. Growing rates are computed as the following:

$$Growing\ Rate = \frac{Nr\ about\ a\ keyword\ in\ the\ LAST\ TWO\ year}{Nr\ about\ ALL fields\ in\ the\ LAST\ TWO\ year} / \frac{Nr\ about\ a\ keyword\ in\ PREVIOUS\ year}{Nr\ about\ ALL fields\ in\ the\ PREVIOUS\ year}$$

(Nr = Number).

2.2 *Overviewing Technologies*

In brief, the analysis is performed by the following steps:

(1) Data of nodes and edges are retrieved from data sources.
(2) A network is built, and the network is clustered.
(3) Labels for each cluster are identified with data of keywords, titles, and abstracts. Depending on the types of intellectual properties, available data types are different.

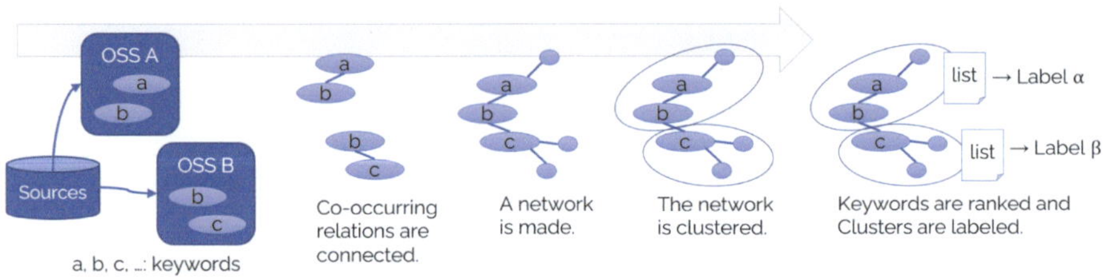

Fig. 7 Steps of clustering a co-occurring keyword network (OSS)

For research analysis, research papers are available as nodes, and citations are edges. Keywords, titles, and abstracts can be used to identify cluster labels. For OSS analysis as shown in Fig. 7, keywords are nodes, and co-existence relations in the same repository are available as edges. Only keywords are available to identify cluster labels.

Looking at all four types of intellectual property, a co-occurring keyword network with keywords as nodes can be used for all intellectual property types, while a citation network can be used for research papers and patents.

2.2.1 Clustering and Scoring Cluster

The co-occurring keyword network is divided into clusters by means of the Newman algorithm (Newman, 2004) for research, Infomap algorithm (Rosvall & Bergstrom, 2008), and the Louvain algorithm (Blondel et al., 2008) for OSS. The reason for the different methods is that the experiments depended on the experiment environments, but the basic idea of the experiment process is the same. The score of a cluster counts up the frequency of keywords by means of integer counting. Then, the clusters are ranked based on the score.

Integer counting (or full counting) (Perianes-Rodriguez et al., 2016) was used. For example, the integer counting regards as one in A and one in B when one OSS has tags of A and B. In contrast, there is fraction counting that regards as half in A and half in B in the previous example. Both integer counting and fraction counting are used as methods for counting research papers in the fields of scientometrics and bibliometrics.

3 Results

The results of applying the methodologies are summarized in Table 2. In the case of research, the emerging keywords and overview showed a consensus conclusion about "perovskite solar cell". This can be considered to be a coincidental encounter

Table 2 List of the results

Methodology	Intellectual property	Data-retrieved year	Section
Emerging keywords for short-term	OSS	2023	3.2.1
Emerging keywords for middle-term	Research	2015	3.1.1
Faded keywords	OSS	2023	3.2.1
Overview	Research		3.1.2
Overview	OSS		3.2.2

at a good time with the consensus conclusion. In the case of OSS, the emerging keywords and overview showed different results.

3.1 Technologies from Research Databases

Research has long been recognized as intellectual property and has been the target of analysis. Researchers in each field will be able to easily detect emerging technologies and grasp the overall picture through daily survey activities in their field. However, it is difficult for people in charge of science and technology policy and intellectual property to observe technologies across disciplines, detect emerging technologies, and grasp the overall picture. The following shows the results of those analyses using data science.

3.1.1 Capturing Middle-Term Emerging Keywords from Research

A part of middle-term emerging keywords in research papers by the previous research (Iwami & Sakata, 2015) are shown in Table 3, with the computed score based on Sect. 2.1.3 and the number of papers was the number of appearances of the keyword in the first web of science category. Keywords related to "perovskite solar cell" were outstanding. After this analysis, the discoverer, Tsutomu Miyasaka, of perovskite solar cell won the 2017 Clarivate Citation Laureates (Clarivate, 2025). Even if different methodologies were used, they could output the same result, "perovskite solar cell". Thus, it can be judged that "perovskite solar cell" are a promising emerging technology.

3.1.2 Overviewing Research

Fortunately, the rapid increase in "perovskite solar cell" can be confirmed not only in the emerging keyword methodology but also in the overview methodology. The top 1% papers published in each year of 2007–2014 are retrieved from the all fields in Web of Science as shown in Fig. 8, and the citation network is divided into clusters.

Table 3 Middle-term emerging keywords

R#	Keyword	Co-occurred Keywords (Computed Score) The First Web of Science Category (Number of Papers)
2	Perovskite solar cell	hole-conductor-free (0.2286); perovskite solar cell (0.1319); organometal halide perovskites (0.1266); ch3nh3pbi3 (0.0903); lead iodide perovskite (0.0845) Materials science multidisciplinary (48)
3	Perovskite solar cells	organometal halide perovskites (0.5127); ch3nh3pbi3 (0.2470) Materials science multidisciplinary (206)
7	Esozoic granites	esozoic granites (0.0860); da hinggan mountains (0.0404); se china (0.0194) Geosciences multidisciplinary (34)
14	Halide perovskite	lead iodide perovskite (0.1972); organometal halide perovskites (0.1234); ch3nh3pbi3 (0.1164); halide perovskite (0.0890) Materials science multidisciplinary (102)
25	Acoustic navigation	acoustic navigation (0.2340); underwater navigation (0.0606); cooperative localization (0.0252); autonomous underwater vehicles (0.0119) Robotics (11)

R# = Rank

Ave. Year is average published year, so larger value means newer domain. "Links" means citations between papers. If the distance between clusters is large, the spring layout (Dengler et al., 1993) will result in a distant relationship. In other words, if there are gaps between clusters in the figure, there is a high possibility that the clusters are separated. Since C#2 is a mixed field, the results of further clustering are shown in Table 4 to answer the interests, although it is not the main topic. As a result, the topic of "perovskite solar cell" was found as one of emerging clusters. The 6th cluster from the top 1% papers has the topic of "organic solar cell". After the cluster about "organic solar cell" divided into sub-clusters, the 5th sub-cluster has the topic about "perovskite solar cell". A surge since 2013 was illustrated in the timeline of "perovskite solar cell", as shown in Fig. 9 (Iwami & Yamashita, 2015).

3.2 *Information Technologies from OSS Repositories*

What has emerged as a new type of intellectual property is something open—open innovation (Chesbrough, 2003), open science (Woelfle et al., 2011), open source software (OSS) (Open Source Initiative, 2021), open data (Auer et al., 2007), open government (OECD, 2021), open source hardware (Pearce, 2016), and so on. Open innovation is a concept of sharing inventions across organizations for better matching individual technologies. It is said that OSS is a successful example of open innovation. As an example of OSS development including open governance activities, a tool for building an information site for coronavirus disease 2019 (COVID-19) was disclosed as OSS and it was also used by some local governments (jgbarah, 2023; Tokyo

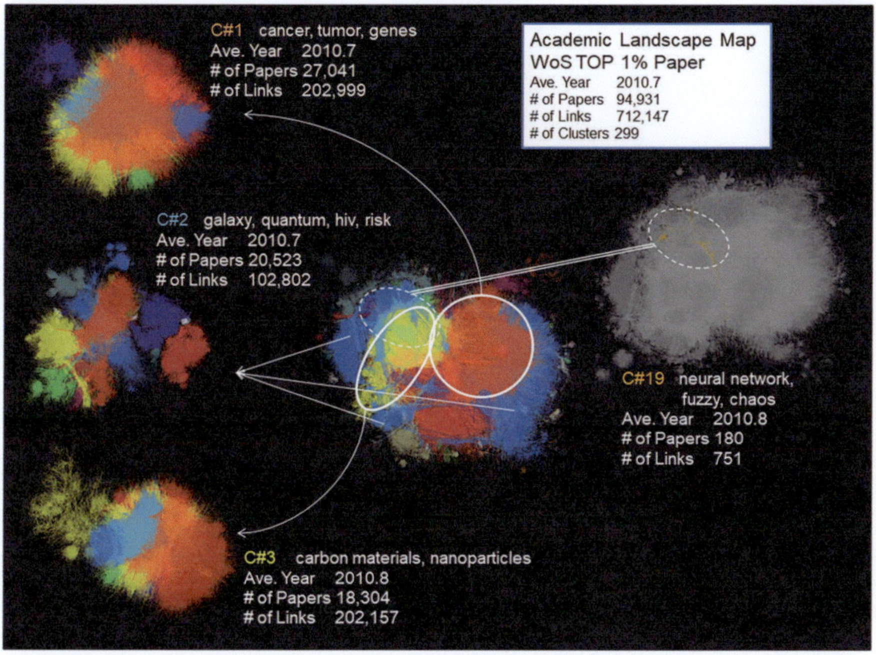

Fig. 8 Citation map of top 1% papers in web of science

Table 4 Breakdown of C#2

C#	Ave. Year	Topic
C#2–1	2010.7	HIV, brain, schizophrenia
C#2–2	2010.6	fuzzy, health, intuition, decision making
C#2–3	2010.7	heart/blood system diseases
C#2–4	2011.0	galaxies, dark matter, Higgs boson
C#2–5	2010.6	quantum spin, topological insulator
C#2–6	2010.7	infection

C# = Cluster Number

Metropolitan Government, 2022). In addition, immediately after the Great East Japan Earthquake, some developers created tools using open data and shared the tool with the people, trying to improve the lives of people during the reconstruction (Kanbara & Shaw, 2021).

Mining software repositories (MSR) activities are being held at several computer science conferences as activities to explore OSS as intellectual property. A particularly large-scale event was the international conference, MSR, where a hackathon was held together with the open source software organization called Community Health Analytics in Open Source Software (CHAOSS) Project (Bitergia, 2023). Several companies have also published analyses of data retrieved from software repositories

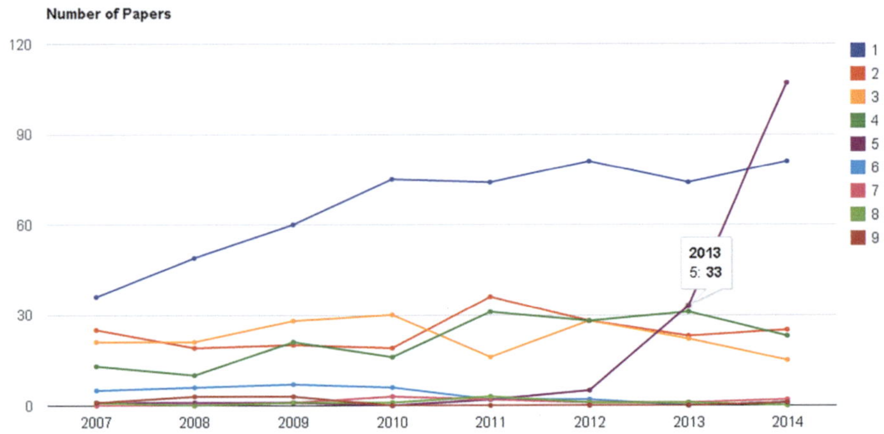

Fig. 9 Timelines of number of papers about "organic solar cell" in web of science

(Iwami, 2022; Mueller & Izquierdo-Cortazar, 2019). OSS has been incorporated into many research and development activities, governments, and companies, and nowadays it has become indispensable in society. In this way, OSS is an area with room for analysis as intellectual property.

3.2.1 Capturing Short-Term Emerging Keywords of OSS

Table 5 shows the emerging keywords that have not appeared in the last three years. The parentheses indicate the number of appearances in that year. The important keywords for security, such as "covid-19" in 2020, "log4j" in 2021, and "ukraine" in 2022, are detected, so this methodology can extract the keywords that need an emergency response even tomorrow. In 2023, keywords related to generative AI have appeared with the excitement of the world. Among the top 20 keywords in 2023, only "pinecone" and "web-ui" are different keywords with AI. The 10th keyword, pinecone, shows a vector database which is strong for AI. The 18th keyword "web-ui" is a wide-applicable word, so it can be related to AI. As a result, all the top 20 keywords in 2023 are related to AI.

From the view to detect unknown topics, the rise in a region can be detected such as dingdong in 2022, which is an online grocery delivery platform launched in Shanghai (Yang, 2023). The descriptions of the repositories related "dingdong" were written in simplified Chinese, so it proves that this methodology can detect software for one region. However, the reason of the detection may be due to invoke many stars following a large number of the large Chinese population. In any case, unlike many other technical terms in rankings, this methodology was able to detect keywords related to services in non-IT industries.

Table 6 shows the faded keywords that have not appeared after they appeared last. Parentheses indicate the number of appearances in that year. These results can be used

Table 5 Short-term emerging keywords on GitHub

	2023	2022	2021	2020
1	(8) Auto-gpt	(9) Diffusion	(9) cve-2021–44,228	(30) Covid-19
2	(7) Alpaca	(5) Ukraine	(7) Log4shell	(17) Coronavirus
3	(7) Voice-conversion	(5) Ding-dong	(7) Log4j	(8) Bugbountytips
4	(6) gpt4all	(4) Chatgpt3	(7) Solana	(8) Github-profile
5	(5) Vicuna	(4) Stable-diffusion-webui	(5) Redteam-tools	(8) Obsidian-md
6	(5) Llmops	(4) Eevanced	(5) Htap	(7) Profile-readme
7	(5) Chatglm	(3) Photo-gallery	(5) Wallet	(7) Obsidian
8	(4) Google-bard	(3) Backup-tool	(5) cve-2021–45,046	(6) Sars-cov-2
9	(4) Dolly	(3) Dreambooth	(4) Clubhouse	(6) 2019-ncov
10	(4) Pinecone	(3) Prompt-engineering	(4) Fiftyone	(6) Zettelkasten
11	(4) Autonomous-agents	(3) Disco-diffusion	(4) Distributed-transactions	(6) Earth-engine
12	(4) Generative-ai	(3) ssl	(4) Generative-adversarial-networks	(6) Nuclei
13	(4) Anthropic	(3) Image-captioning	(4) Bots	(6) Obsidian-plugin
14	(4) Sovits	(3) Vision-language	(4) css-reset	(6) Esbuild
15	(4) Vits	(3) Img2img	(4) cve-2021–45,105	(6) Github-profile-readme
16	(4) So-vits-svc	(3) Txt2img	(4) Tkinter-gui	(5) Covid
17	(4) Audio-analysis	(3) Tcpdump	(3) Code-server	(5) Gaming
18	(4) Web-ui	(3) First-timer-friendly	(3) Libre	(5) wsl2
19	(3) Qlora	(3) Miner-proxy	(3) Bun	(5) Workspace
20	(3) Conversational-ai	(2) Yara	(3) Wsa	(5) Eccv-2020

([digits]): Number of appearances in that year

for withdrawing the fields related to the faded keywords. There are fewer keywords listed in the older years (2017, 2018), and more keywords listed in the younger years (2019, 2020). That means that most of the disappeared keywords could revive again. Different from emerging keywords in Table 5, it is unknown whether the result is a reasonable value. However, it is surprising that broad concepts such as usability in 2017 and www in 2020 are disappearing.

Compared with the top 50 simple keyword counting and triennial co-occurring keyword networks in the Sect. 3.2.2, the selected three keywords (ukraine, log4j, covid) from Table 5 do not appear in the top 50 simple keyword rankings. However, the two keywords (log4j, covid) are included in co-occurring keyword networks as shown in Table 7. It is considered that topics such as "log4j" and "covid" can also be included in co-occurring keyword networks when the topics are largely related to IT. However, the methodology of emerging keywords is better for sudden events like "ukraine". Regarding these keyword detections, the advantage that is different from social news is that it can detect projects that actually use IT to solve problems. However, GitHub is also used for document management and political demonstrations, so it is necessary to check the contents of the project after detecting emerging keywords. If the gaps from social news against this report and existing analyses are acquired, it may be possible to discover new ideas that have not been tackled by IT.

3.2.2 Overviewing OSS

In order to compare with emerging technologies on OSS, results to overview technologies for changes of longer periods (Iwami, 2022) are shown in this section. These results may prove the results of emerging technologies, or they may discover different trends from emerging technologies. Even if an emerging technology is detected, an emerging technology that has a high friendliness with dominant technologies, for example, in terms of language or platform, has a higher chance of gaining more developers and users. Then, the emerging technology may be able to survive in the competitive market. Therefore, it is also important to overview general trends in addition to emerging technologies.

In the methodology of overviewing OSS, keywords are nodes and relations between keywords in the same repository, and then co-occurring keyword networks are clustered. Table 8 shows the ranking by counting keyword appearances before clustering. Table 9 (Infomap algorithm) and Table 10 (Louvain algorithm) show the top keywords within each cluster after dividing co-occurring keyword networks into clusters. The only difference between Table 9 and Table 10 is the clustering algorithm—Infomap algorithm and Louvain algorithm.

In simple counting as shown in Table 8, programming languages are ranked high. In other words, simple counting is difficult to get an overview of the various fields of OSS. Python and JavaScript keep the top two, and React and Go programming language are gradually moving up the ranks. The scores of the keywords "golang" and "go" may be merged. Meanwhile, the significance of clustering is that the activities of the heavily involved OSS and fields are usually seen as having a large impact, but

Table 6 Faded keywords on GitHub

	2020	2019	2018	2017
1	(5) forum	(3) aktienfuehrer	(2) component-driven	(2) add-ons
2	(4) swift-5	(2) bitrise	(2) wxss	(2) usability
3	(3) mock	(2) key-mirror	(2) spoofing	(2) correlation-filters
4	(3) observables	(2) standard-js	(2) nnvm	(2) adversarial-nets
5	(3) markers	(2) snowflake	(2) archived	(2) wasserstein-gan
6	(3) sliding-menu	(2) validate	(2) hello-world	
7	(3) beginners	(2) hapi-server	(2) rxbus	
8	(3) lazy-load	(2) react-native-vector-icons	(2) vault-ui	
9	(2) www	(2) parse-server		
10	(2) regularization	(2) tcomb-form-native		
11	(2) cfw	(2) react-native-router-flux		
12	(2) post-mortem	(2) non-sox		
13	(2) react-book	(2) isometric		
14	(2) elm	(2) depth-maps		
15	(2) o-reilly	(2) convolutional-residual-networks		
16	(2) deep-q-network	(2) source-analysis		
17	(2) code-style	(2) curve		
18	(2) directshow	(2) mojo		
19	(2) rich-text	(2) fps		
20	(2) computer	(2) left-pad		

([digits]): Number of appearances in that year

Table 7 The rank of clusters in co-occurring keyword networks for distinctive keywords

	Louvain algo* in 2020–2022	Louvain algo* in 2017–2019	Infomap algo* in 2020–2022	Infomap algo* in 2017–2019
ukraine	(None)	(None)	(None)	(None)
log4j	(None)	(None)	139	(None)
covid	21	(None)	107	(None)

* algo = algorithm

Table 8 Simple counting of keywords

	2023–2023	2020–2022	2017–2019	2014–2016
1	(1799) python	(2172) python	(1952) javascript	(1360) javascript
2	(1659) javascript	(2116) javascript	(1900) python	(1185) python
3	(1038) react	(1232) react	(1185) android	(699) android
4	(899) golang	(1227) android	(1014) react	(678) java
5	(812) go	(1033) golang	(951) java	(578) php
6	(793) android	(1023) java	(844) golang	(450) go
7	(788) java	(956) machine-learning	(784) go	(443) react
8	(779) typescript	(941) deep-learning	(759) machine-learning	(429) nodejs
9	(734) machine-learning	(915) go	(724) ios	(421) ios
10	(649) deep-learning	(857) typescript	(723) deep-learning	(379) linux
11	(643) php	(766) linux	(721) php	(374) golang
12	(627) linux	(743) ios	(682) nodejs	(339) swift
13	(617) nodejs	(739) nodejs	(636) linux	(332) ruby
14	(598) rust	(738) php	(614) typescript	(293 docker
15	(555) docker	(648) docker	(556) swift	(279) typescript
16	(511) security	(635) rust	(535) docker	(268) c
17	(509) ios	(627) security	(505) vue	(266) awesome
18	(489) kubernetes	(595) vue	(501) security	(265) security
19	(458) vue	(577) pytorch	(478) awesome	(263) css
20	(440) awesome	(554) swift	(433) rust	(259) windows

([digits]): Number of appearances in that period

the clustering overview allows people to see a wide range of other fields. Another major benefit of clustering is that it becomes clear which field of OSS a programming language contributes to.

Comparing the Infomap algorithm (Table 9) and the Louvain algorithm (Table 10), the Informap algorithm obtains clusters with little difference in cluster size, while the

Table 9 Top 20 clusters of OSS with infomap algorithm in 2023

	(Number of Appearances) Top Keywords
1	(15,148) javascript; react; typescript; nodejs; vue
2	(8623) machine-learning; deep-learning; pytorch; data-science
3	(6100) golang; go; java; redis; microservices
4	(4375) docker; kubernetes; aws; devops; containers
5	(3536) linux; macos; windows; cross-platform; electron
6	(2924) python; python3; self-hosted; workflow; python-library
7	(2899) c-plus-plus; game; tutorial; game-development; gamedev
8	(2881) database; sql; mysql; postgresql; mongodb
9	(2822) security; hacking; security-tools; pentesting; osint
10	(2782) cli; terminal; git; shell; command-line
11	(2022) android; kotlin; opensource; android-library; android-development
12	(2018) rust; c; compiler; wasm; webassembly
13	(2010) dotnet; csharp; c-sharp; dotnet-core; architecture
14	(1912) php; laravel; component; symfony; symfony-component
15	(1898) api; graphql; django; rest-api; documentation
16	(1770) video; audio; music; streaming; webrtc
17	(1750) ios; swift; objective-c; xcode; cocoapods
18	(1668) blockchain; ethereum; cryptography; cryptocurrency
19	(1485) monitoring; performance; metrics; logging; prometheus
20	(1424) awesome; awesome-list; list; tools; resources

Louvain algorithm has a large difference in cluster size. Therefore, the Louvain algorithm is efficient for discovering irregular clusters. On the other hand, the Infomap algorithm is suitable when small clusters with high similarity are not absorbed into large clusters. In the both methods, the top three clusters have the same themes: (1) JavaScript, (2) Machine learning, (3) Go programming language, though strictly the third cluster in Louvain algorithm seems to be divided into the third and fourth clusters in Infomap algorithm.

This result was analyzed using an OSS repository, but there are previous studies that analyzed papers as the above Sect. 3.1.2 in the field of software such as OSS (Iwami, 2022). There, the fields “program analysis”, “software process”, “metrics”, “application”, “performance”, “testing”, “source code”, “architecture”, “modeling”, and “developer” were presented. Analysis with the OSS repository in this analysis can provide a concrete overview for practical users, such as programming languages, AI, cloud technology, user interfaces, blockchain, and unmanned aerial vehicles.

Comparing between emerging keywords in the previous Sect. 3.2.1 and the overview in this section, they show different trends. Emerging keyword analysis results have a short-term validity, so they can be used for appealing purposes with rapidly popular keywords. Meanwhile, overview analysis results are long-term

Table 10 Top 20 clusters of OSS with louvain algorithm in 2023

	(Number of Appearances) Top Keywords
1	(18,995) javascript; react; typescript; nodejs; vue
2	(15,405) python; machine-learning; deep-learning; pytorch; data-science
3	(11,073) golang; go; docker; kubernetes; aws
4	(9821) android; linux; ios; macos; windows
5	(7918) rust; c; cpp; c-plus-plus; dotnet
6	(6951) java; database; ruby; sql; mysql
7	(5955) php; api; laravel; http; json
8	(5296) security; privacy; hacking; security-tools; reverse-engineering
9	(4396) cli; terminal; git; shell; vim
10	(3040) awesome; awesome-list; list; algorithms; algorithm
11	(2763) video; audio; music; streaming; image
12	(2450) iot; bot; raspberry-pi; wechat; chat
13	(1569) proxy; network; ssr; vpn; shadowsocks
14	(1302) blockchain; ethereum; cryptocurrency; bitcoin; crypto
15	(1034) chrome-extension; chrome; firefox; crawler; firefox-addon
16	(223) email; smtp; mail; imap; pgp
17	(194) maps; map; openstreetmap; geojson; geo
18	(133) font; fonts; typeface; variable-fonts; ligatures
19	(131) plex; radarr; sonarr; google-drive; emby
20	(106) jekyll; github-pages; jekyll-theme; ruby-gem; jekyll-themes

trends, so they will fit into the investigation of well-used OSS before development and installation.

4 Discussion

4.1 Summary of Results

Although it is impossible to present all combinations of data sources and methodologies in the above results, the middle-term emerging keywords and overviewing results on solar cells match society's subsequent expectations. As of 2024, about 10 years after the analysis, some enterprises have developed them as painting solar cells, wall materials, solar cells for automotive use, and so on (Agency for Natural Resources and Energy, 2024; Yomiuri Shimbun, 2024). This current situation demonstrates that it is possible to detect emerging technologies with high potential for growth.

In addition, short-term emerging keywords captured incidents and rapidly growing trends as COVID-19 and generative AI appeared. This overlaps with topics from social news, but the advantage of this methodology is that it can also detect other top topics. The methodology for short-term emerging keywords is also beneficial.

4.2 General Precautions

When observing keywords over time, there are some cases in which new words are created from existing concepts. Thus, it is noted that keywords about emerging technology are not necessarily really new. This may be due to being taken over from application documents for getting research funding. Alternatively, it may be for ensuring research originality. Meanwhile, "dried[6]" technologies such as water pressure, hydraulic pressure, and pneumatic pressure are key to controlling surgical robots, but pressure has not been detected as an emerging keyword. (It is sure that "surgical robot" can be detected as an emerging keyword.) It should be noted that evidence-based decision-making is not perfect, but it had better to introduce decision-making processes, which is called human-in-the-loop in the testing process.

4.3 Various Ways to Get Keywords

The keywords in this report were prepared in the source database. However, the methodology can be used for any type of keywords. For example, the technology of natural language processing can provide keyword lists even when prepared keywords do not exist. A commonly used method is to extract words from a sentence, but it is necessary to deal with idiomatic problems. Therefore, the prepared keywords provided in a database benefit such analysis. The idiomatic problems are the following:

- Notation fluctuation

    Word transformations (verb tense, noun singular and plural)

- Variation in notation (different words having the same meaning)
- Visually indistinguishable characters: single quote between Windows and Mac

Most of the analysis will be done in English. However, if the analysis is for national policies or technology management of domestic companies in non-English countries, information written in the non-English native language cannot be excluded. If the gap between the English dataset in the world and the dataset written in the non-English native language can be extracted, it will be even a chance to know the gap between the world and each non-English country.

In the case of Japanese which is far from English in language distance, analysts will encounter the following troubles in text analysis:

[6] The "dried" means "fully mature" in the field of information technology.

- Diverse character codes (Representatively, JIS for e-mail, SJIS for Windows, EUC for Linux, UTF-8 for webpages, Base64 encode/decode for URL, and full-width and half-width characters to express katakana in Japanese)
- Alternative characters (New fonts like a character added to the era name when the emperor changed, and old fonts)
- Notation fluctuation (Presence or absence of long sound symbol)

Even when sentences are converted from non-English to English in order to alleviate the problem, problems with word variations arise in the translation process.

4.4 Necessity for Prediction—As the Opposite of the Easiest Way

It would be impossible not to mention artificial intelligence technology as a recent trend. Open models and free server resources have made it easy to tackle natural language processing and prediction. Knowing the future is fascinating, but it doesn't necessarily mean it will actually turn out that way. In addition, people sometimes intentionally realize a future that is different from their predictions. Therefore, we should not be too confident in our predictions. However, predictions can be used for roadmaps and other purposes to promote the development of science and technology in society. People can identify disruptive innovations (Christensen, 1997) that have jumped up from sustaining innovation when the innovations deviate from predictions.

5 Strategies After Detecting Emerging Technologies

Even if some notable results are derived from the quantitative analyses described in the previous sections, further efforts are needed before they can be used for decision-making. It is the same as saying that good technology alone is not enough to generate sales; promotion activities like marketing are also necessary. In technology management, it is sometimes said that data analyses are not useful because of the lack of cooperation between analysis and strategies. Thus, this section describes strategies that decision-makers can utilize to implement the identified emerging technologies.

5.1 Will the Emerging Technology Actually Be Used?

Before touching on how to create and lead emerging technologies, other possibilities are mentioned.

One way is to follow emerging technologies that have been discovered and invented in other organization. Generic drugs are commonly used as old emerging technologies. Generic drugs are manufactured cheaply with the same ingredients as the original after the patent expires (Reiffen & Ward, 2005). The advantage is that the demand and production process can be estimated. The disadvantage is the opportunity by getting a higher market share at the early stage is missed, and in the case of generic drugs, that the cost is required to get government approval.

Another way is to get the entire company that owns the technology through mergers and acquisitions. The great advantage of this is that you can introduce not only the technology, but also skilled personnel, an established system, and revenue outlook. However, the disadvantage is that the acquisition costs are high. For example, Skype is a video calling system created in Estonia, and it was acquired by Microsoft in 2011 (Microsoft, 2024a). Skype's technology was absorbed into Microsoft's technology, and Skype for business use will end in 2021 (Microsoft, 2024b). It is unfortunate that the original technology is absorbed and the name disappears, but it functions as an ecosystem of innovation that allows creators to try new technologies with the money from selling their companies and to invest resources in the next emerging technology.

In the next section, examples of using emerging technologies, which have been discovered and invented by oneself, are introduced.

5.2 *What Happened After Emerging Technologies Had Appeared?*

By utilizing the methodologies proposed in this paper, promising emerging technologies will be found and those technologies will be adopted. What should you do after that? Research and development investments may not be recovered if the adopted technology does not gain market share and provide a return worthy of continuous development. In other words, it is necessary to compete with similar technologies and expand market share.

As a case of the famous failure to occupy the market, the functions of wristwatches, MP3[7] players, digital cameras, mobile phones, and personal digital assistants (PDAs) were absorbed into the functions of smartphones. To be precise, it was successful up to a certain point, but later it could no longer stand on its own. Just before its integration into smartphones, the iPod worked with a software distribution system called iTunes, giving it an edge over the hardware-only MP3 player (Kotama, 2018). Sony has made innovations that expand the environment in which hardware can be used from indoors to outdoors, with the concept "taking your music with you and enjoying it easily" by "walkman" (Haire, 2023). Apple has expanded functionalities and contents in addition to the innovation of Sony.

In the results of emerging sciences (Sect. 3.1), perovskite solar cells were discovered as an emerging technology. Then, what happened after that? Perovskite solar

[7] MP3 means "MPEG-1 Audio Layer-3". However, the MP3 player is not called the "MPEG-1 Audio Layer-3 player".

cells announced in Japan continued to be developed in Japan. However, the first discoverer and Japanese research institutes did not exclude others from this promising technology. Such open development is an effective strategy for expanding technology (PatSeer, 2022), but for business open development may give too much of an advantage to competitors.

5.3 *Alliance Strategies*

As mentioned in the previous section, while there is competition among companies, even greater competition is occurring. In competitions between economic blocs (Kop, 2023) and competitions between corporate groups, the bigger ones capture the market. Therefore, in order to become bigger, alliance strategies around emerging technologies have become popular.

Several strategic movements can be observed in the OSS community. In the OSS community, organizations don't just participate in software development to make the product advantageous for their company. There is a movement to launch foundations for specific technologies, to partner with other organizations, and to make the technologies mainstream. In one case, after Google addressed SLSA (Supply chain Levels for Software Artifacts) with its own cloud service, it is promoting the SLSA via one foundation "OpenSSF (Open Source Security Foundation)" in the OSS community in order to make it a required standard for users and other organizations (Linux Foundation, 2023a). In the competition for cloud services, Google has yet to be able to catch up with the top two companies, Amazon and Microsoft, but it is guessed that its efforts in issuing SLSA are aiming to acquire customers who require high quality rather than gaining market share. Meanwhile, the Overture Maps Foundation started without Apple and Google which are providing the map service with the overwhelming number of users (Overture Maps Foundation, 2023).

Below, a case of open alliance strategies and a case of closed alliance strategies on recent emerging technologies are introduced.

5.3.1 A Case of Open Alliance Strategy

In recent years, in the rapidly growing field of artificial intelligence (AI), key machine learning frameworks have emerged such as Deeplearning4j (Deeplearning4j, 2023), Keras (Keras, 2023), TensorFlow (TensorFlow, 2023), PyTorch, and others. At one time, TensorFlow incorporated Keras, and was the top share framework. However, around 2018 or 2019 the share of PyTorch exceeded that of TensorFlow in research paper Implementations grouped by machine learning framework (META AI, 2023), and around 2021 the same thing occurred in the Google Trend as shown in Fig. 10. In 2022, Meta (the former Facebook), the developer of PyTorch, transferred it as OSS to the neutral Linux Foundation, and the PyTorch Foundation was launched (The Linux Foundation, 2023b). Google, the main developer of TensorFlow, is also participating

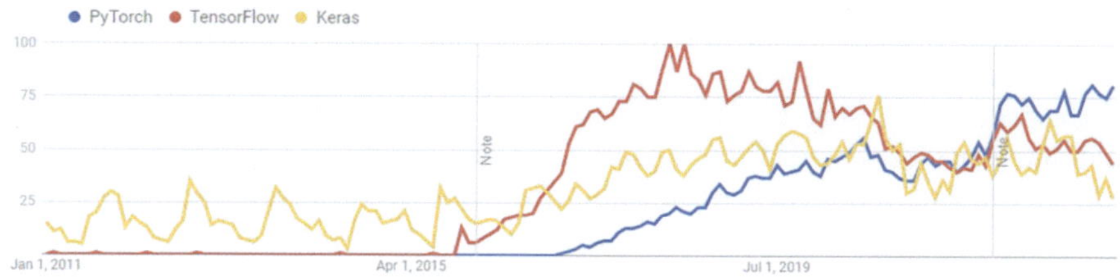

Fig. 10 Timelines of PyTorch, TensorFlow, and Keras on Google Trends (Worldwide, 1/1/2011–6/6/2023, Computers & Electronics)

in the foundation. As of 2023, PyTorch keeps the superiority against TensorFlow, and it appears that PyTorch is heading toward becoming the dominant player. Behind such a situation of the competition of machine learning frameworks, in the AI-embedded system development, there has been a shift from an era of development that starts from creating models to an era of development using provided models, and HuggingFace has increased its presence in providing models. PyTorch may not be the definitive solution yet at present.

5.3.2 A Case of Closed Alliance Strategy

Regarding generative AI, a different alliance strategy from machine learning frameworks is adopted. This is the traditional alliance strategy and it would be understandable without "closed". Generation AI was roughly a competition between ChatGPT series (OpenAI, 2023) and Bard/Gemini series (Google, 2023) around November 2023. BART incorporated Transformer in 2018. The ChatGPT developer, OpenAI, collaborated with many companies and made ChatGPT a major trend. On the other hand, Google's Bard is a laggard and is not as popular as ChatGPT. Note that neither is OSS, and only the interface is publicly available. In other words, it can be said that they adopted the closed strategy to expand alliances. However, there is an expert who published the prediction that OSS will overtake both during the competition between ChatGPT and Bard (Patel & Ahmad, 2023). In 2024, Meta's Llama series (Meta, 2024), which has been made OSS, and Anthropic's Claude series, which temporarily surpassed the performance of the ChatGPT series (Anthropic, 2024), are also joining the fierce development competition. As for generative AI, the end point regarding the best strategy is still not in sight.

5.3.3 Applications Outside the Software Field

Is alliance strategy possible in areas other than software? Nowadays, cars are no longer just mechanical engineering but have computers inside them, so several car companies are participating in collaborative OSS projects about embedded software (for example, Automotive Grade Linux). OSS projects are also being implemented in fields such as energy (for example, LF Energy), agriculture (for example, AgStack) (Linux Foundation, 2023c), and medicine (for example, Nightscout Foundation) (Nightscout Foundation, 2023). The process of introducing IT and software into all fields has made it easier to utilize alliance strategies via OSS communities. In order to search for alliance partners in open innovation, the classification of alliance partners (Shaikh & Levina, 2019) will be supportive in utilizing alliance strategies.

6 Conclusion

This report provided the easiest methodology to detect emerging technologies and indicate the effects of the methodology. Comparing the emerging technologies, the overviews of technologies were also shown. Additionally, from the cases of emerging technologies and productization, the strategies to expand the market share of the technologies are important for the technologies to survive. Alliance strategies to increase presence in the market were introduced.

Finally, both emerging technologies and the data science that detects them are wonderful innovations that foretell a convenient society in the future. However, society is a place for humans, and we should keep in mind that it is humans who will lead promising science, technology, and innovation in the correct direction.

References

Adner, R., & Levinthal, D. A. (2002). The emergence of emerging technologies. *California Management Review, 45*, 50–66. https://doi.org/10.2307/41166153

Agency for Natural Resources and Energy: Perovskite solar cells: the key to expanding renewable energy in Japan (Part 2)—Efforts aimed at early social implementation (In Japanese), https://www.enecho.meti.go.jp/about/special/johoteikyo/perovskite_solar_cell_02.html, last accessed 2024/05/29.

Ajanovic, A. (2015). The future of electric vehicles: Prospects and impediments. *Wires Energy and Environment, 4*, 521–536. https://doi.org/10.1002/wene.160

AI Alliance, https://thealliance.ai/, last accessed 2024/05/30.

Anthropic: Introducing the next generation of Claude, https://www.anthropic.com/news/claude-3-family, last accessed 2024/05/30.

Kodama, M. (2018). Apple versus Sony: strategy transformation by capabilities congruence through asset orchestration. In: Sustainable Growth Through Strategic Innovation. pp. 103–129. Edward Elgar Publishing. https://doi.org/10.4337/9781785366383.00010.

Auer, S., Bizer, C., Kobilarov, G., Lehmann, J., Cyganiak, R., & Ives, Z. (2007). DBpedia: a nucleus for a web of open data. *Presented at the*. https://doi.org/10.1007/978-3-540-76298-0_52

Barredo Arrieta, A., Díaz-Rodríguez, N., Del Ser, J., Bennetot, A., Tabik, S., Barbado, A., Garcia, S., Gil-Lopez, S., Molina, D., Benjamins, R., Chatila, R., & Herrera, F. (2020). Explainable artificial intelligence (XAI): concepts, taxonomies, opportunities and challenges toward responsible AI. *Inf. Fusion., 58*, 82–115. https://doi.org/10.1016/j.inffus.2019.12.012

Bitergia: GrimoireLab—Software Development and Community Analytics platform, https://chaoss.github.io/grimoirelab/, last accessed 2023/05/28.

Blondel, V. D., Guillaume, J.-L., Lambiotte, R., & Lefebvre, E. (2008). Fast unfolding of communities in large networks. *Journal of Statistical Mechanics: Theory and Experiment, 2008*, P10008. https://doi.org/10.1088/1742-5468/2008/10/p10008

Bonvoisin, J., Mies, R., Boujut, J.-F., Stark, R. (2017). What is the "Source" of Open Source Hardware? *Journal of Open Hardware. 1*, (2017). https://doi.org/10.5334/joh.7.

Bradford, S. Serendipity, happenstance, and luck: the making of a molecular tool. https://www.the-scientist.com/serendipity-happenstance-and-luck-the-making-of-a-molecular-tool-71497, last accessed 2024/05/30.

Chan, C. M. L. (2013) From open data to open innovation strategies: creating E-Services using open government data. *In: 2013 46th Hawaii International Conference on System Sciences.* pp. 1890–1899. IEEE. https://doi.org/10.1109/HICSS.2013.236.

Chapelle, F. H. (2014). The history and practice of peer review. *Groundwater, 52*, 1–1. https://doi.org/10.1111/gwat.12139

Chesbrough, H. W. (2003). *Open Innovation: The new imperative for creating and profiting from technology*. Harvard Business School Press.

Christensen, C. (1997). The innovator's dilemma. *Harvard Business Review Press.*

Clarivate: Identifying extreme impact in research, Clarivate Analytics uses citations to forecast Nobel Prize winners, https://ir.clarivate.com/news-events/press-releases/news-details/2017/Identifying-extreme-impact-in-research-Clarivate-Analytics-uses-citations-to-forecast-Nobel-Prize-winners/default.aspx, last accessed 2025/12/18.

Copeland, S. (2019). On serendipity in science: Discovery at the intersection of chance and wisdom. *Synthese, 196*, 2385–2406. https://doi.org/10.1007/s11229-017-1544-3

Deeplearning4j, https://deeplearning4j.konduit.ai/, last accessed 2023/10/31.

Dengler, E., Friedell, M., Marks, J. Constraint-driven diagram layout. *In: Proceedings 1993 IEEE Symposium on Visual Languages*. pp. 330–335. IEEE Comput. Soc. Press. https://doi.org/10.1109/VL.1993.269619.

Engineering and Technology History Wiki: IEEE Standards Association History, https://ethw.org/IEEE_Standards_Association_History, last accessed 2023/10/16.

Fusco, S. (2019). Lessons from the past: the venetian republic's tailoring of patent protection to the characteristics of the invention. *SSRN Electronic Journal*. https://doi.org/10.2139/ssrn.3331687

Gartner: Gartner Hype Cycle, https://www.gartner.com/en/research/methodologies/gartner-hype-cycle, last accessed 2024/05/28.

Gest, H. (1997). Serendipity in scientific discovery: a closer look. *Perspectives in Biology and Medicine, 41*, 21–28. https://doi.org/10.1353/pbm.1997.0008

GitHub: GitHub REST API, https://docs.github.com/en/rest, last accessed 2023/05/28.

Goedhuys, M., & Sleuwaegen, L. (2013). The impact of international standards certification on the performance of firms in less developed countries. *World Development, 47*, 87–101. https://doi.org/10.1016/j.worlddev.2013.02.014

Google: Bard - Chat Based AI Tool from Google, Powered by PaLM 2, https://bard.google.com/, last accessed 2023/10/29.

Haire, M.: A Brief History of the Walkman, https://content.time.com/time/nation/article/0,8599,1907884,00.html, last accessed 2023/10/31.

Hoffmann, M., Nagle, F., & Zhou, Y. (2024). The value of open source software. *SSRN Electronic Journal*. https://doi.org/10.2139/ssrn.4693148

IEC: History, https://www.iec.ch/history, last accessed 2023/10/06.

Iwami, S., Yamashita, Y. (2015). Comparative bibliographic analyses of the global and Japanese databases in the field of solar cell. *In: 2015 IEEE International Conference on Industrial Engineering and Engineering Management (IEEM)*. pp. 1198–1202. IEEE. https://doi.org/10.1109/IEEM.2015.7385837.

Iwami, S., Sakata, I. (2015). Growing topics and emerging researches in science. *In: The International Society for Professional Innovation Management (ISPIM)* (2015).

Iwami, S. (2022). A methodology to categorize rapidly changing projects of open source software. *In: 2022 Portland International Conference on Management of Engineering and Technology (PICMET)*. pp. 1–8. IEEE. https://doi.org/10.23919/PICMET53225.2022.9882581.
Jeevan, H., & Ranganathan, S. (2004). A new basis for the classification of quasicrystals. *Journal of Non-Crystalline Solids, 334–335*, 184–189. https://doi.org/10.1016/j.jnoncrysol.2003.11.035
jgbarah: Free, open source software for COVID contact tracing, https://jgbarah.github.io/Notes/foss-covid-tools.html, last accessed 2023/10/30.
Kanbara, S., & Shaw, R. (2021). Disaster risk reduction regime in Japan: an analysis in the perspective of open data. *Open Governance. Sustainability., 14*, 19. https://doi.org/10.3390/su14010019
Keras: Deep Learning for humans, https://keras.io, last accessed 2023/10/31.
Kop, M. Democratic countries should form a strategic tech alliance, https://ssrn.com/abstract=3814409, last accessed 2023/10/30.
Kronick, D. A. (1990). Peer review in 18th-century scientific journalism. *JAMA, the Journal of the American Medical Association, 263*, 1321. https://doi.org/10.1001/jama.1990.03440100021002
Le Gall, D. (1991). MPEG. *Communications of the ACM, 34*, 46–58. https://doi.org/10.1145/103085.103090
Leydesdorff, L., Milojević, S. (2012) Scientometrics.
Leydesdorff, L., Kushnir, D., & Rafols, I. (2014). Interactive overlay maps for US patent (USPTO) data based on International Patent Classification (IPC). *Scientometrics, 98*, 1583–1599. https://doi.org/10.1007/s11192-012-0923-2
Lieberman, M. B., & Montgomery, D. B. (1988). First-mover advantages. *Strategic Management Journal, 9*, 41–58. https://doi.org/10.1002/smj.4250090706
Longo, L., Brcic, M., Cabitza, F., Choi, J., Confalonieri, R., Ser, J. Del, Guidotti, R., Hayashi, Y., Herrera, F., Holzinger, A., Jiang, R., Khosravi, H., Lecue, F., Malgieri, G., Páez, A., Samek, W., Schneider, J., Speith, T., Stumpf, S. (2024). Explainable artificial intelligence (XAI) 2.0: a manifesto of open challenges and interdisciplinary research directions. Inf. Fusion. 106, 102301 (2024). https://doi.org/10.1016/j.inffus.2024.102301.
META AI: Papers With Code : Trends, https://paperswithcode.com/trends, last accessed 2023/10/30.
Mathew, G., Agrawal, A., & Menzies, T. (2023). Finding trends in software research. *IEEE Transactions on Software Engineering, 49*, 1397–1410. https://doi.org/10.1109/TSE.2018.2870388
Meta: Meta Llama, https://llama.meta.com/, last accessed 2024/05/30.
Microsoft (2024a). Microsoft Officially Welcomes Skype, https://news.microsoft.com/2011/10/13/microsoft-officially-welcomes-skype/, last accessed 2024/07/23.
Microsoft (2024b). Upgrade from Skype for Business Online to Microsoft Teams in Microsoft 365, https://www.microsoft.com/en-us/microsoft-365/previous-versions/skype-for-business-online, last accessed 2024/07/23.
Mueller, D., & Izquierdo-Cortazar, D. (2019). From art to science: the evolution of community development. *IEEE Software, 36*, 23–28. https://doi.org/10.1109/MS.2019.2936177
Newman, M. (2004). Fast algorithm for detecting community structure in networks. *Physical Review E, 69*, Article 066133. https://doi.org/10.1103/PhysRevE.69.066133
Nightscout foundation: the nightscout foundation, https://www.nightscoutfoundation.org/, last accessed 2023/10/29.
Nobel Prize Outreach: The Sveriges Riksbank Prize in Economic Sciences in Memory of Alfred Nobel 1987, https://www.nobelprize.org/prizes/economic-sciences/1987/summary/, last accessed 2023/10/06.
OECD: Open government, https://www.oecd.org/gov/open-government/, last accessed 2021/08/19.
OpenAI: ChatGPT, https://openai.com/chatgpt, last accessed 2023/10/29.
OpenChain: Welcome to the OpenChain Project—OpenChain, https://www.openchainproject.org/, last accessed 2023/10/22.

Open Source Security Foundation: Open Source Security Foundation (OpenSSF) Best Practices badge, https://www.bestpractices.dev/en, last accessed 2023/10/06.

Open Source Initiative: The Open Source Definition, https://opensource.org/docs/definition.php, last accessed 2021/08/19.

Overture Maps Foundation: Overture Maps Foundation, https://overturemaps.org/, last accessed 2023/10/28.

Patel, D., Ahmad, A. Google "We Have No Moat, And Neither Does OpenAI," https://www.semianalysis.com/p/google-we-have-no-moat-and-neither, last accessed 2023/10/29.

PatSeer: Patent Portfolio Analysis of the Key Companies in Perovskite Solar Cells. (2022).

Pearce, J. M. (2015). Quantifying the value of open source hardware development. *Modern Economy, 6*, 1–11.

Pearce, J. M. (2016). Return on investment for open source scientific hardware development. *Sci. Public Policy., 43*, 192–195. https://doi.org/10.1093/scipol/scv034

Perianes-Rodriguez, A., Waltman, L., & van Eck, N. J. (2016). Constructing bibliometric networks: A comparison between full and fractional counting. *Journal of Informetrics., 10*, 1178–1195. https://doi.org/10.1016/j.joi.2016.10.006

Raymond, E. S. (2000). The cathedral and the bazaar.

Reiffen, D., & Ward, M. R. (2005). Generic drug industry dynamics. *Review of Economics and Statistics, 87*, 37–49. https://doi.org/10.1162/0034653053327694

Rogers, E. (2003). Diffusion of Innovations. Simon and Schuster.

Rosvall, M., & Bergstrom, C. T. (2008). Maps of random walks on complex networks reveal community structure. *Proceedings of the National Academy of Sciences, 105*, 1118–1123. https://doi.org/10.1073/pnas.0706851105

Rotolo, D., Hicks, D., & Martin, B. R. (2015). What is an emerging technology? *Research Policy, 44*, 1827–1843. https://doi.org/10.1016/j.respol.2015.06.006

Shaikh, M., & Levina, N. (2019). Selecting an open innovation community as an alliance partner: Looking for healthy communities and ecosystems. *Research Policy, 48*, Article 103766. https://doi.org/10.1016/j.respol.2019.03.011

Shimizu, H. (2019). Innovation going wild: overcoming the "Lost 20 Years" of the Japanese economy (In Japanese). *Shinchosha* (2019).

Solow, R. M. (1956). A contribution to the theory of economic growth. *The Quarterly Journal of Economics, 70*, 65. https://doi.org/10.2307/1884513

Suarez, F. F., Grodal, S., & Gotsopoulos, A. (2015). Perfect timing? Dominant category, dominant design, and the window of opportunity for firm entry. *Strategic Management Journal, 36*, 437–448. https://doi.org/10.1002/smj.2225

Tarver, E. First mover: what it means, examples, and first mover advantages, https://www.investopedia.com/terms/f/firstmover.asp, last accessed 2023/10/06.

Teich, A. H. (2018). In Search of Evidence-based Science Policy: From the Endless Frontier to SciSIP. *Annals of Science and Technology Policy. 2*, 75–199. https://doi.org/10.1561/110.00000007

TensorFlow, https://www.tensorflow.org/, last accessed 2023/02/02.

Tokyo Metropolitan Government: Tokyo COVID-19 Task Force website, https://github.com/tokyo-metropolitan-gov/covid19, last accessed 2022/02/23.

Tranchard, S.: ISO celebrates 70 years, https://www.iso.org/news/2017/02/Ref2163.html, last accessed 2023/10/06.

Tucker, L.: 10 of the oldest universities in the world, https://www.topuniversities.com/blog/10-oldest-universities-world, last accessed 2023/10/06.

The Linux Foundation (2023a). OpenSSF Announces SLSA Version 1.0 Release, https://openssf.org/press-release/2023/04/19/openssf-announces-slsa-version-1-0-release/, last accessed 2023/10/28.

The Linux Foundation (2023b). PyTorch Foundation, https://pytorch.org/foundation, last accessed 2023/05/31.

The Linux Foundation (2023c). Linux Foundation. Browse Projects, https://www.linuxfoundation.org/projects, last accessed 2023/10/29.
The Linux Foundation (2024a). LF AI & Data, https://lfaidata.foundation/, last accessed 2024/05/30.
The Linux Foundation (2024b). CNCF Landscape, https://landscape.cncf.io/stats, last accessed 2024/05/28.
The Yomiuri Shimbun: Toyota, Kyoto Startup to Partner on Bendable Solar Cells for EVs, https://japannews.yomiuri.co.jp/business/companies/20230628-119199/, last accessed 2024/05/29.
von Hippel, E., & von Krogh, G. (2003). Open source software and the "private-collective" innovation model: issues for organization science. *Organization Science, 14*, 209–223. https://doi.org/10.1287/orsc.14.2.209.14992
Wellman, B. (2001). Computer networks as social networks. *Science* (80-.). 293, 2031–2034 (2001). https://doi.org/10.1126/science.1065547.
Woelfle, M., Olliaro, P., & Todd, M. H. (2011). Open science is a research accelerator. *Nature Chemistry, 3*, 745–748. https://doi.org/10.1038/nchem.1149
Yang, Z. Shanghai's lockdown is giving China's online grocery apps a second chance, https://www.technologyreview.com/2022/04/29/1051604/shanghai-lockdown-online-grocery/, last accessed 2023/05/31.

Weak Signal Detection in Quantitative Foresight: A New Method for Large-Scale Data Using Large Language Models

Arash Hajikhani

1 Introduction

The increasing speed of technology adoption with the introduction of groundbreaking technologies such as deep learning (LeCun et al., 2015), cryptocurrencies (Nakamoto, 2008), and mRNA vaccines (Sahin et al., 2014), the ability to anticipate future trends has become increasingly vital. Technological change and adoption underscoring the critical importance of anticipation (Flyverbom, & Garsten, 2021) and strategic foresight (Rohrbeck & Kum, 2018). "Future-making," a proactive approach to future (Wenzel, 2022), is pivotal in maintaining strategic resilience, identifying emerging business opportunities (Moor, 2005; Yoon, 2012), and cultivating competitive advantage (Rohrbeck & Kum, 2018).

Research on even predictable surprises highlight that organizations remain ill prepared to respond to even the foreseeable events, (Grant & Wunder, 2021). This realization has led to calls for vigilance and investment in foresight capabilities (Schoemaker & Day, 2021), which would allow for detecting not just "strong signals"—trends that are readily apparent from past data—but also "weak signals," those subtle indicators that may identify significant future shifts. The conventional methods, largely based on expert opinions (Ebadi et al., 2022; Sutherland & Woodroof, 2009), are not devoid of biases, calling attention to the need for more comprehensive and robust methodologies to better understand the state of the art and accurately capture the dynamics of weak signals, which are crucial for robust foresight creation.

Addressing these concerns, this research introduces a novel approach for the identification of weak signals through extensive data analysis. It critically examines the existing literature on the identification of weak signals in science and technology, pinpointing several key shortcomings. Traditional methodologies often struggle with

A. Hajikhani (✉)
VTT Technical Research Centre of Finland, Espoo, Finland
e-mail: Arash.Hajikhani@vtt.fi

H. da Hora et al. (eds.), *Technology Mining*,
https://doi.org/10.1007/978-3-032-10849-4_5

the uncertainty and inherent risks in the identification of technological innovation, thus hindering the accurate identification of nascent trends and their intensities (Guo et al., 2012). Additionally, conventional approaches have a limited scope, usually concentrating on individual technologies and failing to offer a holistic analysis and comparison of multiple technological advances (Lee et al., 2011). This limitation restricts a comprehensive understanding and forecasting of emergent trends. Moreover, the challenges associated with big data, including the complexities of diversity, reduction, integration, indexing, querying, and analysis, are crucial for effectively comprehending and interpreting vast data quantities to identify weak signals in science and technology, which also lead to extensive manual work (Rotolo et al., 2015). These challenges necessitate a balanced and flexible approach that simplifies large-scale data while ensuring the retention of significant signals and trends, and also reducing manual effort in the process. Large-scale datasets are distinguished by their substantial volume, diversity, and complexity, necessitating specialized tools and techniques for effective processing and analysis. The wide range of topics within these datasets requires advanced computational methods to manage, process, and analyze the vast and varied collection of records. Utilizing sophisticated tools designed for large-scale data is essential to handle the challenges posed by such extensive datasets.

This study utilizes large language models to facilitate the identification of key terms and their subsequent distillation into signals from any form of temporal textual data. We define weak signals based on the set of characteristics outlined by Hiltunen (2008) and develop a novel quantification of these characteristics. Hiltunen (2008) proposed three dimensions for future signs: signal, indicating visibility; issue, indicating the diffusion of the future sign; and interpretation, representing the meaning of the future sign to the receiver. A weak signal exhibits low levels of signal, issue, and interpretation. To avoid potential subjectivity in this three-dimensional framework, Yoon (2012) suggested that a proper quantification of the dimensions would be needed (Yoon, 2012).

Weak signals exhibit evolutionary behaviors compared to other signals, underpinned by two primary constructs: analyzing the signal of the signals and their issue (Yoon, 2012). Yoon (2012) constructs indicators for these dimensions as the "emergence" of the signal and the "issue" of the signal. However, the construction of Yoon's indicators is vulnerable to bias depending on the nature and structure of the data to which it is applied. Significantly, the indicators are vulnerable to data in which the number of documents varies across the selected time periods, and cases in which a signal exhibits erratic growth trend. Our study presents modified, and where necessary, new indicators addressing these challenges and providing a solution that can be more widely applied for large-scale data analysis and unequal data populations. Using these indicators, weak signals are identified and categorized, then subjected to field expert reviews for sanity checks and validation. While this study involves expert review for the validation of signals, the use of large language models (LLMs) for identification, categorization, and clustering tasks can reduce the dependency on manual sanity checks during the signal analysis process.

In this study, we operationalize the proposed methodology using scientific publication data in the context of water networks. By 2025, it is projected that 3.5 billion people will reside in water-scarce areas, with water demand anticipated to increase by 30% by 2050 (United Nations, 2019). This scenario compels water utilities and industries to seek innovative solutions for water efficiency and sustainable management (Idrica, 2023). The water sector's role in climate adaptation is increasingly vital, necessitating the scaling up of technologies such as water reuse and smart irrigation. The industry is also experiencing growth in investment, focusing on sustainability and digitalization. This trend presents substantial opportunities for technology companies to develop and introduce relevant solutions. Moreover, technology firms must remain vigilant to environmental changes to sustain competitiveness (Reeves & Deimler, 2011).

Research institutes and universities are utilizing weak signals to guide future research directions and establish collaborative partnerships, enhancing their potential for industrial application and funding. Key areas of development among utilities include reducing leaks in water distribution networks, lowering energy consumption in water infrastructure, and broadly aiming for decarbonization and the circular economy (Frost & Sullivan, 2022). While some leading water utilities are already achieving net-zero operations, new technologies and solutions are essential to address further challenges in the operational environment, induced by factors such as climate change and rapid urbanization.

The theoretical contribution of this research lies in developing a scalable, quantitative foresight framework using LLMs to identify weak signals. This reduces reliance on manual analysis and broadens the scope of input data. The methodology facilitates multi-perspective weak signal detection across contexts and contextualizes signals with greater granularity. Adaptable to various data sources and repeatable over time, this approach allows continuous observation of emerging trends and reduces dependency on field experts.

In terms of practical implications, this research provides significant insights for managers, researchers, and policymakers by highlighting the value of integrating weak signals from various sources to identify new business opportunities and potential disruptions. By adopting cross-disciplinary studies and advanced methods, organizations can enhance their foresight and make informed decisions, crucial for long-term success.

The remainder of this chapter is structured as follows: Sect. 2 provides a comprehensive review of existing theories. Section 3 details the data sources and the methodology employed for data processing and analytics. Section 4 presents the research findings. Section 5 discusses these findings, elucidating both the theoretical and managerial implications of this study, while also highlighting its limitations and suggesting avenues for future research.

2 Background

Foresight and Weak Signals

As a strategic management discourse, foresight traces its roots to the early 1900s when Nobel laureate Alfred North Whitehead first recognized the need for developing "*philosophical competencies to understand the complexity of societies*" (Rohrbeck, Battistella & Huizing, 2015). In more contemporary terms, the growing complexity of the business environment called for capabilities that allowed organizations to "*see through the apparent confusion, to spot developments before they become trends, to see patterns before they fully emerge, and to grasp the relevant features of social currents that are likely to shape the direction of future events*" (Tsoukas & Shepherd, 2004). In the century following its conceptual inception, foresight has taken leaps in theory, methodology and practical application toward a more purpose-driven paradigm. Today, the strategic management literature regards foresight as a dynamic capability that allows businesses to create a better competitive position in future markets vis-á-vis their competitors (Schwarz, Rohrbeck & Wach, 2020; Semke & Tiberius, 2020; Marinkovic et al., 2022; Rohrbeck & Kum, 2018; Ahuja et al., 2005; Hamel & Pralahad, 1994). Future competitive advantage is created as foresight allows firms to "*identify, observe and interpret factors that induce change, determine possible organization-specific implications, and trigger appropriate organizational responses [..., allowing] access to critical resources ahead of competition, preparing the organization for change, and permitting the organization to steer proactively towards a desired future*" (Rohrbeck et al., 2015).

Foresight elevates an organization's capabilities to acquire, process, learn from and exploit information in the form of future signals (Baskarada et al., 2016). This translates into, for instance, anticipatory governance (Flyverbom, & Garsten, 2021), first mover advantages, strategic agility to respond successfully to external changes (Vecchiato, 2015), into the capabilities to influence others and an enhanced capacity for organizational learning to identified benefits (Rohrbeck & Schwarz, 2013). In more concrete terms, businesses that robustly employ foresight and future-making practices in their operations have been found—more often than their peers with less robust foresight approaches—to gain insights into the changes of the business environment, to have an enhanced understanding of customer needs, to identify potential customers, to have an enhanced understanding of markets, to identify opportunities and threats for their offering, to create the ability to adopt alternative perspectives, to improve the coordination of business objectives, and to anticipate so-called wild cards earlier (Buder, 2020; Wenzel, 2022). Foresight has been found to further support business renewal as it can provide insight on key questions such as those around technology selection, future customer needs and potential disruptions (Calof et al., 2018). Finally, to quantify the value these capabilities imbue on business, Rohrbeck and Kum (2018) have shown that companies that strategically employ foresight display 33% higher profitability and 200% higher market capitalization rates than their less future-oriented competition.

As the value of foresight in strategic management is steadily becoming clearer, the motivation to develop evermore effective foresight methodology and processes is growing in parallel. For simplicity's sake, the process of foresight can be split into two phases; detection of pre-cursors of potential change and making sense of how such changes affect the state of the world in the form of opportunities and threats to business (Marinkovic et al., 2022). Both phases necessitate their respective methodological approaches.

Aligned with the primary objective of this study, we focus on a methodology for the early detection of trends and shifts, a key phase addressed by weak signal theory (Martinsons et al., 2001). The concept of weak signals, pioneered by Ansoff (1975), plays a crucial role in identifying early symptoms of strategic discontinuities—significant departures from past trends leading to strategic surprises for organizations. These signals serve as early warnings of potential threats or opportunities, initially vague but gradually becoming clearer, informing the firm's strategic response (Ansoff, 1975; Ansoff et al., 2018). As the signals develop, they require a range of analytical techniques, from expert opinion to quantitative modelling, depending on the clarity of the information. This progression in knowledge informs the strategic responses of firms, categorized by Ansoff into external and internal adjustments to enhance awareness, increase flexibility, or directly address emerging threats or opportunities (Ansoff, 1975, 1980, 1985; Ansoff et al., 2018). Ansoff also emphasizes the strategic benefit of preparedness, enabling organizations to respond more swiftly as signals clarify, thus reducing response times and leveraging early warnings for strategic advantage (Ansoff, 1975; Moorman & Miner, 1998).

Subsequent research on weak signals has sought to refine Ansoff's initial concepts and explore their implications for management. Efforts have been made to provide more detailed definitions, differentiate between signals and the phenomena they indicate, and discuss the importance of signaled phenomena (Mendonça et al., 2004). This discussion has broadened to include the identification and interpretation of weak signals, stressing the importance of a collective approach to detect these subtle indicators of future change (Hiltunen, 2001). Highlighting the importance of unconventional sources and overcoming paradigm blindness suggests that incorporating a broad spectrum of inputs, from expert opinions to "expert amateurs," is vital for effective weak signal detection. This collective sensitivity, driven by diverse perspectives, is deemed essential for organizations to effectively navigate strategic discontinuities and surprises (Day & Schoemaker, 2004). Weak signals, characterized as subtle yet indicative anomalies, predict potential changes (Xue et al., 2019). Challenges such as cognitive biases and the absence of systematic methodologies hinder weak signal detection, as noted by Popper (2008). In response, structured frameworks are required for better classification and analysis of weak signals (Kaivo-oja & Lauraéus, 2018; Cory, 2023). By employing advanced techniques to robustly utilize large-scale quantitative data in particular, identification of weak signals could avoid issues of cognitive bias and lack of systematic approaches, filling gaps in understanding future trends and trajectories and enhancing the reliability and accuracy of projected scenarios.

Application of Data and Methods in Identification of Weak Signals

The utilization of data and methodologies in identifying and analyzing weak signals and foresight in technology transcends traditional boundaries, incorporating technological, social, economic, and environmental facets. This multidisciplinary approach involves a range of empirical techniques, such as indicators and trend analysis, citation analysis, co-word analysis, overlay mapping, and patent analysis (Bates et al., 2015). These methods are essential in assessing innovation, tracking technological progress, and forecasting future technologies (Rotolo et al., 2015). In scenarios with limited historical data, alternative strategies like available large-scale data such as patent trend extrapolation and bibliometric analysis are used to predict burgeoning technological domains (Daim et al., 2007). Notably, patent and bibliometric analysis is valued for its ability to quantitatively assess technological risks and opportunities in high-tech sectors, particularly through the analysis of knowledge flows (Benoit et al., 2018).

The integration of AI-based data mining models in recent years, highlighted in the work of Mühlroth and Grottke (2020), has significantly improved the automation and precision in detecting and forecasting technological trends in large-scale scientometrics analytics (Xu et al., 2019). Ha et al. (2023) introduce a methodology employing machine learning-based approaches in automating identification of weak signals. This advancement underscores the evolving role of AI and Language Learning Models in this field which provides a multifaceted view of potential technological paths and opportunities (Wang & Zhang, 2020; Wu et al., 2021).

Beyond technological methodologies, foresight studies are increasingly adopting systemic approaches, integrating network analysis and scenario planning. This holistic perspective assists in understanding the evolution of social, economic, and environmental systems (Wiebe et al., 2018). These studies combine methodologies from various disciplines, including sociology, political science, economics, management, and cultural studies, providing a comprehensive outlook on future scenarios. For instance, the social shaping of technology (SST) framework is used in green technology foresight to explore environmental potentials and risks, especially in nano, bio, and information and communication technologies (Mao et al., 2020).

The emergence of large language models (LLMs) and advanced natural language processing (NLP) techniques has also significantly influenced the field of emerging signals and foresight. LLMs, particularly those based on neural networks and transformer architectures, have demonstrated remarkable efficiency in identifying emergent topics by analyzing linguistic patterns and detecting anomalies, thanks to their few-shot learning capabilities (Kim et al., 2019; Ting et al., 2012). These models, exemplified by the transformative impact of tools like ChatGPT, have been applied across various sectors, including healthcare, social media, and finance (Liu et al., 2016), enhancing productivity and addressing complex tasks while also necessitating a focus on their limitations and ethical implications (Dwivedi et al., 2023).

Nonetheless, enhancing the robustness and transparency of these models remains a priority (Bradley et al., 2007).

The integration of advanced Natural Language Processing (NLP) techniques and Large Language Models (LLMs) has substantially enhanced the effectiveness of technology foresight and weak signal analysis. This combination offers invaluable insights into emerging technologies and their potential impacts by efficiently processing and interpreting vast amounts of textual data. LLMs can analyze and understand complex patterns in text, enabling more accurate predictions and deeper insights into technology trends. They can handle large-scale data analysis by quickly sifting through extensive textual information, identifying relevant trends, and synthesizing this information into understandable insights, thus aiding in more informed decision-making processes.

In conclusion, the literature motivates a need for more research in strategic management's field of foresight, particularly in the early detection of weak signals. Such endeavors are pivotal, as effective foresight is crucial for organizations to maintain competitive advantages, as evidenced by higher profitability and market capitalization of firms adept in foresight. Therefore, this research tackles this need by proposing a methodology focused on the robust analysis of large-scale data, integrating the advanced capabilities of LLMs. This approach aims to overcome the cognitive biases and systematic approach deficiencies, enhancing the accuracy and reliability of foresight. By leveraging cutting-edge technologies, the research fills a critical need for more sophisticated and reliable tools in strategic management, aligning with the evolving requirements of effective foresight in an increasingly complex business environment.

3 Data & Methods

This section describes the data and methodology used to investigate trend emergence in the context of water management and water networks. First, we discuss the sourcing of scientific publications from Web of Science (WoS) and detail the data processing and construction. Then we describe the methodology of analysis, which builds heavily on natural language processing and the utilization of LLMs.

Data

The data collected for this study comprise a set of publications within the domains of water management, water networks, water distribution, and smart water applications. This data was obtained from the Web of Science (WoS) platform, and the data collection was executed on August 21, 2023.

The search strategy employed WoS's Topic Search (TS) field, which covers the fields of title, abstract, author keywords, and WoS's "Keywords Plus". The search terms included the following queries: "water" within a three-word radius of "network*," "water" within a three-word radius of "distribution," "water monitor*," "water system*," "water management," and "smart water."[1] This yielded a total of 83,504 publications, published since the year 2010.

The bibliographic data obtained through this process included metadata associated with each publication. This contained details such as the publication date, title, abstract, authorship, keywords, citation counts, publication type, and more. This study does not employ the author reported keywords or the WoS-generated "Keywords Plus" associated with the publications for its keyword analysis. This is for two reasons, first is to avoid bias that may be introduced from authors' self-defined keywords, and second is to avoid introducing selection bias based on the publications for which these fields are available, as both fields exhibited a high rate of missing values (Ebadi et al., 2022).

The construction of the final dataset necessitates preliminary data cleaning procedures. Publications that lacked essential attributes such as title, abstract, or publication year were excluded from the dataset. To facilitate keyword extraction, the titles and abstracts were concatenated to construct a unified text field. This resulted in a dataset of 81,357 publications spanning the years 2010 to 2023.

Figure 1 shows the count of publications in this dataset over time, stacked by publication type. Publications in the fields of water management, water networks, water distribution, and smart water applications have grown rapidly in the past 10 years. This is mainly driven by growth in journal articles, which make up the majority of publications each year and an increasing portion of the total publications over time. The number of conference papers in these fields has remained relatively constant. The count of series publications is less than 10 per year. The overall count in 2023 is low as the data was retrieved in August of 2023.

Method

The methodology delineated in this study augments previously established frameworks for signal detection to more flexibly and efficiently extract emerging topics from large-scale data. The quantification of weak signal characteristics has previously been applied by Ebadi et al., (2022) and Yoon (2012), building on characteristics of weak signals established by Hiltunen (2008). Our method selectively expands and modifies Yoon's quantification of what Hiltunen defined as signals' "signal" and "issue". Our quantitative construction of these dimensions differs from Yoon's, so we define our constructions using descriptive terms for these characteristics: specifically, we define quantitative constructs for diffusion (representing

[1] We use the following query to retrieve the publications on water network context:

TS = (((water NEAR/3 network) OR (water NEAR/3 distribution) OR "water monitor" OR "water system" OR "water management" OR "smart water")).

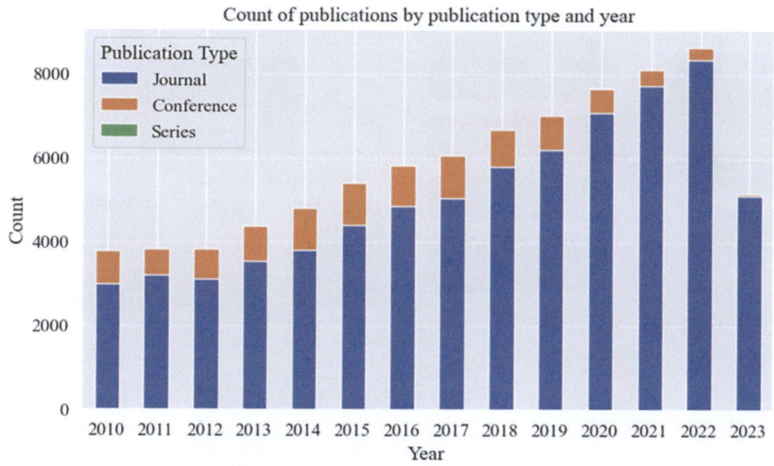

Fig. 1 Count of publications by publication type and year

Hiltunen's characteristic of "issue") and intensity (representing Hiltunen's characteristic of "signal"). Our constructions represent these dimensions accurately over large-scale, non-uniform data, allowing for greater flexibility across data contexts. Figure 2 displays a high-level overview of the data processing and analysis. The methodology depicted in the figure initiates with the initial data processing and time segmentation. This is followed by keyword extraction and processing. The subsequent phase concentrates on signal extraction, which is based on the diffusion and intensity measures. Once the signals are identified, the methodology advances to the categorization and validation stages.

Time Segmentation

After initial processing, as described in Sect. 3.1, the dataset contains 81,357 publications from 2010 to 2023. Next, the data is segmented into time intervals for improved signal detection and time periods for later temporal comparison. Yoon (2012) suggests signals be identified within specific time intervals, with alterations in signal behavior across these intervals indicating the degree of emergence for the signal. The temporal comparison across multiple periods aids in discerning how the signals have evolved or diminished throughout the study duration. In addition, observing the activity of signals detected in the most recent period allows for better identification of current emergence trends.

After experimentation with different time interval lengths, we chose to identify the signals in two-year intervals. This was deemed the most appropriate interval length to capture keyword representation and coverage across the full data. This resulted in seven time intervals, where the final time interval [2022, 2023] represents a slightly

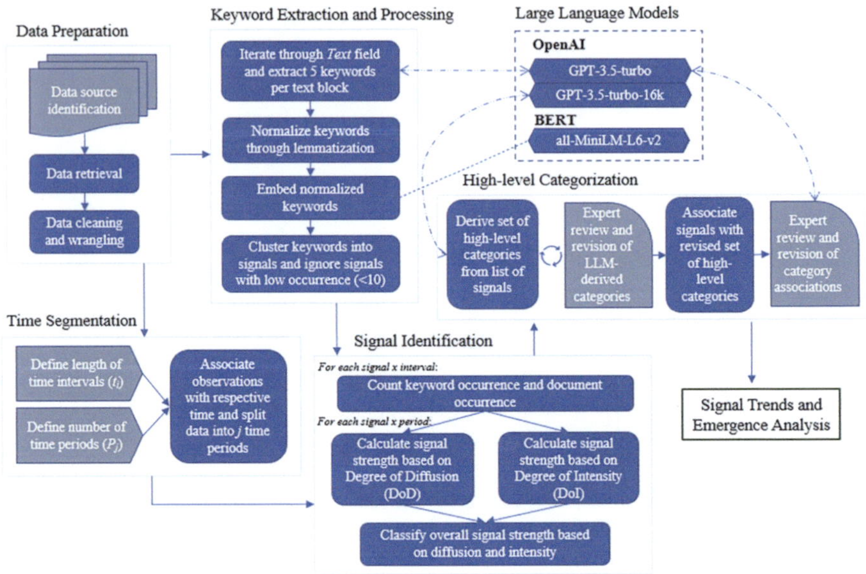

Fig. 2 Conceptual flow of the signal detection analysis

shorter time frame, since 2023 is captured only through August. The time intervals are referred to as t_1 to t_7 hereon.

Based on the seven two-year time intervals, we further divided the data into two periods. The periods are referred to as P_1 and P_2 and cover the following time intervals:

- Period 1 (P_1): [2010, 2011] to [2014, 2015]; (t_1 to t_3)
- Period 2 (P_2): [2016, 2017] to [2022, 2023]; (t_4 to t_7)

Keyword Extraction via LLM

The process of keyword extraction leverages recent advancements in natural language processing (NLP) techniques and large language models (LLM). As previously discussed, this approach can mitigate potential biases stemming from authors' keyword selections. We also avoid using pre-defined set of keywords by experts for the following reasons: (1) These predefined collections possess a limited number of keywords. (2) The keywords may exhibit the experts' biases, or lack of awareness about certain emerging trends. (3) The granularity of the keywords may vary across different publications. Instead, employing LLMs enables the extraction of a more comprehensive set of keywords, taking into account the context of the surrounding information. LLMs are advanced machine learning architectures, primarily rooted in deep learning. They are designed to understand and analyse human-like text through training on vast amounts of textual data, capturing intricate linguistic patterns and nuances. Ebadi et al., (2022), introduced the use of BERT for selecting keywords from text. Bidirectional Encoder Representations from Transformers (BERT) is a family of language models developed by researchers at Google (Devlin et al., 2018). BERT models have been widely leveraged and have represented the state of the art for contextualized keyword extraction in recent years.

In this research, we employ various large language models to enhance the efficiency and precision of keyword extraction. The advent of generative pre-trained transformer (GPT) models, exemplified by OpenAI's GPT series, has ushered in novel capabilities and methods for linguistic analysis and representation.

In the selection of a suitable LLM for the keyword extraction task, BERT models and OpenAI's GPT 3.5 model were considered. The BERT-based models were utilized via the KeyBERT library, a well-established Python package for keyword extraction. After testing multiple configurations, it was determined that OpenAI's GPT-3.5-turbo model returned superior keywords with respect to: (1) preventing the selection of general keywords; (2) keywords' contextualization and significance to the given text; (3) flexibility in the selection of n-grams considering the text context. The keyword extraction was then performed over each publication's title and abstract pairing to select the 5 main keywords of the text through API calls to OpenAI's Chat Completion endpoint, utilizing the GPT-3.5-turbo model.

For the majority of the publications, the LLM successfully returned five keywords as instructed. However, for 13 publications, OpenAI's "harmful content" filter was triggered, and no keywords were returned. For example, in the case of a paper on the topic of water management during genocidal political regimes. For an additional 414 publications, 2 to 17 (excluding 5) keywords were returned. The majority of these cases returned six keywords (339 publications) or seven keywords (37 publications).

Keywords Selection

While keyword extraction using LLMs offers the advantage of considering the context of the surrounding discussion, it also introduces more variations in the keywords. These variations can result in keywords with high semantic similarities being treated as distinct. Consequently, these variations can introduce biases in the final interpretation of signals. For instance, a signal might manifest in multiple formats with a high degree of similarity yet be treated as different. To illustrate, consider the case of keywords such as "cathode current collectors," "current collector foils," "cathodic current collector," and "cathode current collector." Ebadi et al., (2022) attempted to mitigate this problem by manually examining the extracted keywords to select a final list of keywords. Still, this manual examination can be significantly labor-intensive, especially when analyzing large-scale data, and may also overlook keywords with minor variations in format, potentially omitting significant information. To counteract the impact of variation in keyword format, in this study we cluster similar keywords. This approach not only reduces the number of unique keywords but also ensures the comprehensive capture of information conveyed by signals that appear in slightly different linguistic forms.

An efficient algorithm is essential for clustering large-scale datasets within an acceptable timeframe. In this study, we utilized the Sentence Transformers model for advanced text embeddings. Notably, SBERT (Sentence Embeddings using Siamese BERT-Networks) supports embeddings in over 100 languages (Reimers & Gurevych, 2019). These embeddings facilitate comparisons, for instance, using cosine similarity, to identify sentences with analogous meanings. This is pivotal for tasks such as semantic textual similarity, semantic search, and paraphrase mining. Although traditional agglomerative clustering algorithms are often sluggish for vast datasets, SBERT presents a rapid clustering solution, capable of handling large datasets, called community detection. Our methodology sets a specific threshold for cosine similarity, beyond which two keywords are deemed similar enough to belong to the same community, or cluster. Through iterative testing, we determined 0.75 to be the appropriate cosine similarity threshold for this research. After clustering the similar keywords, we remove clusters with occurrence rates too low to be analysed within the dataset.

Signal Analysis

After extracting and grouping keywords into clusters that represent signals, we compute two metrics to characterize signal strength, representing the signals' intensity and diffusion. The intensity measure gauges the average growth rate of signals' prevalence during a given period. The diffusion measure evaluates the growth rate of signals' appearance across documents. Our study introduces several substantial modifications to the metrics introduced by Yoon (2012) and used by Ebadi et al. (2022). First, we streamline the measures by excluding the time-weighted components used in previous work, aiming for a uniform evaluation across all years. This is necessary due to the bias that the time-weighted components can introduce. Specifically, a high value driven by these time-weighted parameters could potentially offset a decline in the intensity or diffusion of a signal, misrepresenting it as a positive growth rate. Second, for the denominator of the intensity metric, we determine the intensity of keyword i in time interval j by averaging over the number of documents that mention the signal, instead of averaging over the total number of documents published in the given time interval. This adjustment allows for improved differentiation between the intensity and diffusion measures. Yoon's (2012) approach determines visibility through an average over the document count in each interval and is designed for consistency. However, this leaves the metric vulnerable to bias if the document counts differ significantly across intervals. Our method mitigates this issue by adjusting for variations in document numbers across intervals to ensure a balanced representation of intensities. Third, we refine the diffusion metric by eliminating the division by total number of documents, as used in previous work. Figure 1 illustrates that publications related to the water network context have roughly tripled in recent years. Consequently, incorporating the total number of publications into the equation could skew the ratio, particularly for weak signals that were infrequently observed during their early emergence.

The following formulas present the intensity and diffusion measures for a given keyword in a certain time interval.

For all time intervals t over all periods P, calculate the degree of intensity ($DoI_{i,j}$) and the degree of diffusion ($DoD_{i,j}$) of each keyword:

$$DoI_{i,j} = \left(\frac{TF_{i,j}}{DF_{ij}}\right). \quad (1)$$

$$DoD_{i,j} = DF_{i,j}. \quad (2)$$

where TF_{ij} is total keyword frequency of keyword i in time interval j and DF_{ij} is document frequency of keyword i in time interval j.

After calculating $DoI_{i,j}$ and $DoD_{i,j}$ for each keyword, we compute the average growth rate (geometric mean) of each signal over each period. These growth rates are then used to construct a Keyword Intensity Map (KIM) and a Keyword Diffusion Map (KDM) for every period. On the KIM, the x-axis represents the average (arithmetic mean) *DoI* across the time intervals in the period, while the y-axis depicts the *DoI* growth rate (geometric mean). For the KDM, the x-axis signifies the average document frequency of the signals, and the y-axis represents the *DoD* growth rate. As illustrated in Fig. 3, each map is partitioned into four quadrants based on the median values of the x and y axes.

Yoon (2012) categorizes signals as latent, weak, well-known, and strong. As our research focuses on emerging trends, we instead select signals identified as either weak or strong on both maps. This classification yields four distinct cohorts: "nascent" (signals positioned in the weak quadrant on both maps), "concentrated" (signals categorized as strong on the intensity map but weak on the diffusion map),

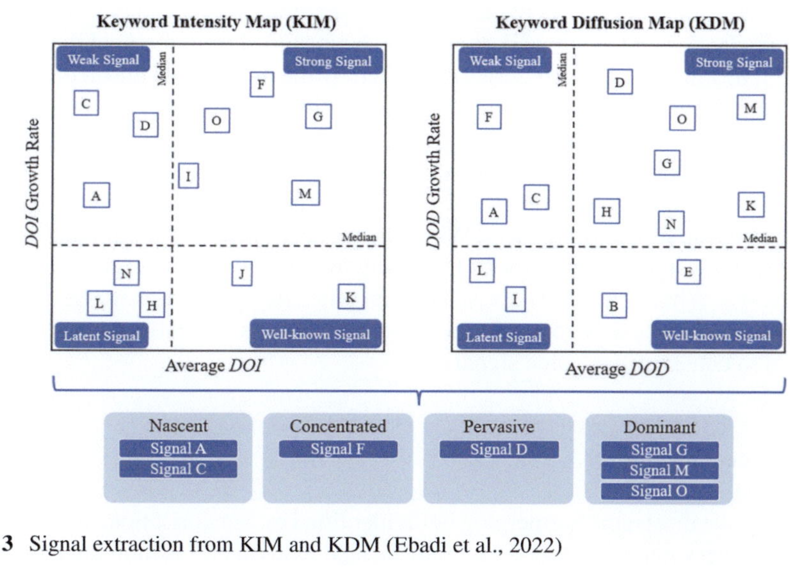

Fig. 3 Signal extraction from KIM and KDM (Ebadi et al., 2022)

"pervasive" (signals are positioned as weak on the intensity map but strong on the diffusion map), and lastly "dominant" signals (signals positioned in the strong quadrant on both maps).

Moreover, we only consider the signals showing positive growth in both indices from the first to the second period. This implies that we assign scores to the signal ranging from minimum to maximum, corresponding to its progression from latent, to weak, and then to well-known, and finally, to strong.

Categorisation and Validation

In large-scale quantitative approaches, we may end up with a vast number of signals, which necessitate presentation of the signals at a higher level of granularity for a comprehensive understanding. Previous research typically employed field experts for such procedures (Ebadi et al., 2022; Yoon, 2012). While we acknowledge that experts offer precision, this process is highly labor intensive. In this research, we leverage large language models to assist signal categorization, thereby streamlining the expert review of the identified signals.

The process initiates with deriving main topics conveyed by the signals. The full body of signals is given as input to the LLM, which returns a set of unique, high level categories representing themes detected in the list of signals. By reiterating this experiment, the reviewing expert can recognize recurrent themes in the categories to determine which categories are most representative. If multiple iterations return a similar category, for example, "data modelling," "data analysis," "modelling and simulation," "simulation," "analytics and modelling," this suggests that theme, in this case data modelling and analytics, is significant within the signals and a common category can be derived. Conversely, if "agricultural methods" appears in one iteration with no similar categories in the remaining iterations, this category is considered less representative of the signal data and may be excluded from the final list of categories. In this manner, the categories suggested by the model are subjected to expert review and refined into a final set. This final set of categories is then deployed in the subsequent step for categorization of the signals using LLM, which assigns each signal to one or more of the categories.

After the category assignment via the LLM, expert review is again utilized for validation of the signal assignment. When the number of signals at this stage is very high, assigning signals to categories first using the LLM generates significant time–cost savings. While this allows for substantial efficiency gains, we strongly suggest subsequent expert review to verify these assignments, as specialized knowledge is required to contextualize some signals.

Emergence Analysis

To identify the most rapidly emerging fields identified by categorisation, the decision-making process must incorporate a multi-faceted assessment of individual signals.

This involves not only considering the quantity of such signals but also evaluating their degree of emergence. This assessment begins with first ranking each signal's degree of emergence using a cumulative distribution function. The ranking is then normalized on a scale from zero to one, relative to the signal's counterparts within the same cohort. Subsequently, the aggregation of the normalized rankings, initially at the individual signal level (average of ranking of the signals constructs) and subsequently at the broader field level, provides a quantifiable measure of the degree of emergence within each field. Equation 3 shows the procedure used to measure the degree of emergence for each category. The subsequent summation of these combined scores within the category then quantitatively indicates the degree of emergence for each category.

$$Degree\ of\ Emergence = \sum mean(Intensity\_rank, Diffiusion\_rank) \tag{3}$$

4 Results

The keyword selection process in this study identified a total of 170,975 distinct keywords. Among these, "water management," "water quality," and "climate change" were the top three most frequent keywords, with occurrences of 4,701, 1,807, and 1,801, respectively. As previously discussed, keyword sparsity can significantly impact the final aggregation of signals. To mitigate this issue, we applied SBERT clustering, which resulted in 75,434 distinct clusters. After testing various thresholds for minimum similarity within a cluster, we ascertained 0.75 to be an appropriate threshold. However, among all these clusters, clusters with very low total keyword occurrence were excluded, leaving 4,708 clusters in our sample. The number of keywords in each cluster ranges from 1 to 677. The average number of keywords in each cluster is 9 keywords, and 90% of the clusters consist of 29 keywords or fewer. Clusters are named by the keyword with the highest occurrence rate. The three most frequent clusters were the "water management' cluster with 8,453 occurrences, the "irrigation" cluster with 3,730 occurrences, and the "water quality" cluster with 3,609 occurrences. Additionally, it was observed that 16 clusters emerged exclusively in the second period of the study. This clustering process allowed for the analysis to group keywords with similar meaning and relevancy into what will henceforth be referred to as signals.

After calculation of $DoI_{i,j}$ and $DoD_{i,j}$, we mapped the signals on to the KIM and KDM quadrants. To evaluate the emerging signals, we keep only the signals allocated to one of the four cohorts (nascent, concentrated, pervasive and dominants) with growth in the behaviour from period 1 to period 2. This process yielded 342 signals. Following an expert review, the list was refined to include 295 signals, categorized as 82 nascent, 47 concentrated, 63 pervasive, and 103 dominant signals. It is important to note that the exclusion of certain signals may vary depending on the study's context.

Additionally, potential noise in the initial dataset could be attributed to the selection criteria used in the data retrieval query, which might influence the final list.

To categorize the signals, we begin with a review of the main topics discussed in the list of selected keywords. Utilizing the OpenAI GPT-3.5-16k model, we initially identified a list of topics including (1) water quality and contamination, (2) environmental impact and sustainability, (3) agricultural systems and practices, (4) climate and weather patterns, (5) energy and power systems, (6) health and disease, (7) infrastructure and engineering, (8) modelling and simulation, (9) nanotechnology and materials science, and (10) remote sensing and data analysis. However, following expert review, we arrived at four areas of interest to emerging technologies and data topics in the water network literature. The categories were defined as (1) analytics and modelling, (2) data, (3) technology, and (4) application areas. The final assignment of signals to these four areas was conducted first using the OpenAI GPT 3.5 model, followed again by three experts review. The experts' suggestions were then compared and any disagreements of signal assignment to a category between two of the experts were settled with the third expert. After expert review, the number of signals requiring modified classification in each category was compared against the initial classifications returned by the LLM. Comparing the results of LLM assignments with the expert review revealed the validity rate of the initial LLM assignment across the final set of categories. Table 1 displays validation metrics representing the performance of the AI in assigning signals to topical categories, relative to the final expert approved assignments.

Table 1 shows validation metrics for the AI category assignment of signals against expert reviewer's assignments. Column 2, "ERs assignments / AI assignments," represents the ratio of signals the expert reviewers assigned to the given category to signals the AI assigned to the given category. Column 3, "Valid AI assignments / Total AI assignments," represents the rate at which an AI's assignment of a signal was accurate, with accuracy defined as the expert reviewers' assignment of the signals. As can be seen in column 2, the AI was much more conservative in its assignments of signals to a category than the expert reviewers were, with the expert reviewers assigning on average four times the number of signals to a category as the AI. Column 3 shows that for the "Analytics and Modelling" and "Data" categories,

Table 1 Validation of AI category assignments against expert review

Category	ERs assignments/AI assignments	Valid AI assignments/total AI assignments
(1)	(2)	(3) (%)
Analytics and modelling	2.58	94.7
Data	3.64	81.8
Technology	7.17	66.7
Application area	3.91	45.5

*ERs = Expert Reviewers; AI = OpenAI GPT 3.5-Turbo LLM (Nov 2023)

the AI's categorization, when it did assign a signal to a category, was fairly accurate. Conversely, for "Technology" and "Application Area," the performance was comparatively lower. Overall, while the AI assigned signals more conservatively than the expert reviewers, its accuracy of assignment appears relatively higher for more explicit categories, while relatively lower for categories with broader or more vague interpretations. This demonstrates the importance of developing well-defined categories in the previous step, as well as thorough model prompting. Additionally, the interdisciplinarity of the signals may influence the accuracy of the model's assignment, as well. Note that this validity rate may vary depending on the context of the case. For example, the validity may be lower for cases in which the context is very specialized. Such cases could result in more nuanced or niche categories more likely to require specialized knowledge for accurate classification. Likewise, for less specialized cases, the validity may be higher.

The final analysis after the categorization delineates the four primary categories collectively encompass 132 signals. It is important to clarify that the allocation of signals to the "other" category does not diminish their significance. Rather, it reflects the experts' focus on the four categories of interest. For instance, there are several signals under the "Other" category relate to socio-demographic factors, such as "population trends," "socio-economic development" and economic dispatch," which directly impact water consumption and the need for development of efficient water management technologies. Such keywords have not appeared in the four categories focused on for this case.

The findings show that 45% of the signals were deemed relevant to the four categories of interest. Of the relevant signals, 41 are characterized as nascent, 16 correspond to concentrated signals, 42 have been identified as pervasive, and 33 are classified as dominant signals.

As shown in Fig. 4, nascent signals are more prominently observed in the analytics and modelling category, while the technology category contains the highest number of dominant signals.

Concerning the emergent analysis, as can be seen in Fig. 5, the analytics and modelling category contains the highest degree of emergent among nascent signals. However, the technology category leads with respect to the presence of dominant signals in the water network. The results mirror the currently prevailing trends in the water sector, which include the adoption of digital water technologies and implementing AI to analyze data to detect patterns and identify areas of inefficiency. However, while the benefits and necessity of obtaining on-line data on the water network has been generally acknowledged, the availability of reliable robust sensors have largely remained uncertain. This implies that in many utilities, data collection is only setting off and consequential analytics and modelling not yet developed into full maturity.

At the same time, water utilities are facing challenges with rapid changes and uncertainties in their environment relating to extreme weather events and local water scarcity. This circumstance creates new needs for forecasting, which, in turn, contributes to the growing interest in analytics and modelling. Though employment

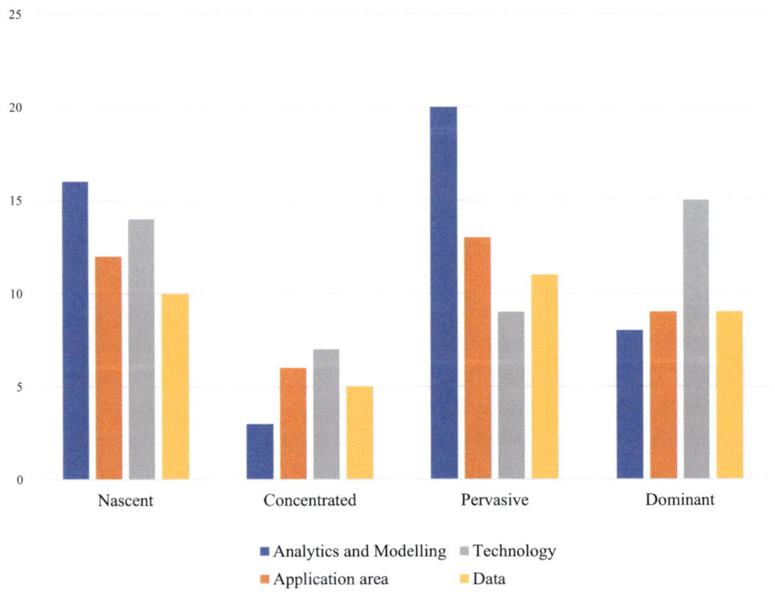

Fig. 4 Count of signals by category

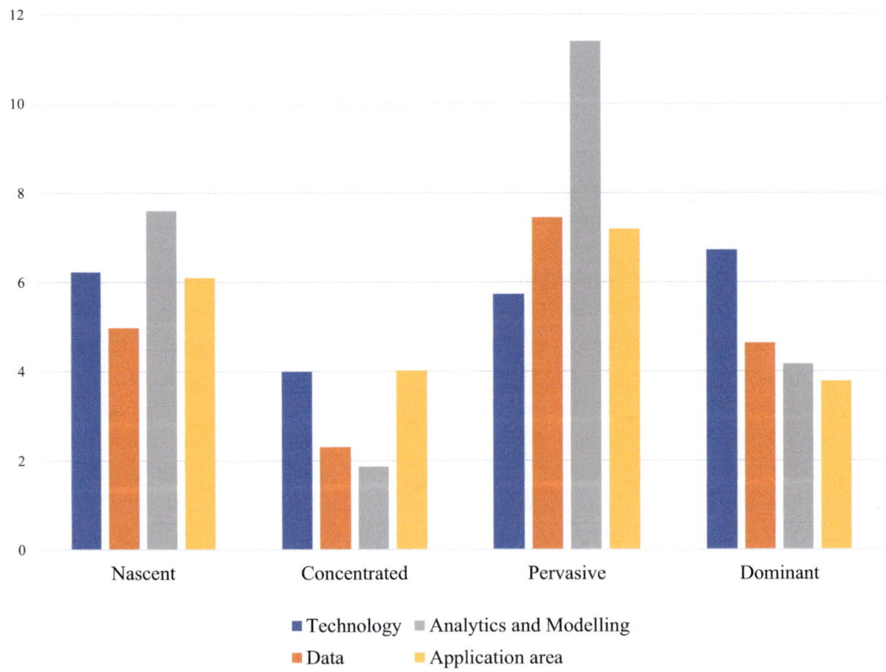

Fig. 5 Degree of emergence for the selected areas by category

will require large investments in digital capabilities, this trend is foreseen to only strengthen in the years to come (von Ditfurth et al., 2021).

5 Discussion and Conclusion

This research introduces a novel methodology for identifying weak signals based on extensive textual corpora using large language models. While weak signals are challenging to identify in their early stages, they become much easier to recognize over time (Ha et al., 2023). The approach centers on the identification of signals based on the scholarly well-established features of weak signals in terms of diffusion and intensity. The advent of large language models has significantly enhanced textual analysis, facilitating quantitative foresight by reducing manual efforts. While previous studies have developed novel methods for detecting weak signals in textual data, these methods also often require extensive manual input from field experts, limiting their application to large-scale data and niche contexts (Ebadi et al., 2022; Ha et al., 2023; Yoon, 2012).

Our study leverages the capabilities of large language models to streamline the identification of weak signals, thereby reducing reliance on manual analysis and broadening the scope of input data. Building upon the well-established characteristics of weak signals offered by Hiltunen (2008), our methodology offers large-scale quantitative approach in foresight literature. Generating insights using rigorous methods on a comprehensive dataset allows researchers to integrate the results with in-depth qualitative methods, leading to a more complete understanding of future scenarios (Karlsen, 2014).

Moreover, categorizing identified signals within specific areas aids in evaluating their degree of emergence in context. Our method quantifies this degree of emergence, acknowledging that the adoption of certain technologies may vary across fields and over time. A technology well-established in one context could still be emerging in another. This approach advances previous studies such as Li et al. (2021) that have attempted to gauge the degree of technological emergence on a general level using large-scale data.

Deploying the methodology in the water network context assisted in comprehending the various segments of water infrastructure. As the literature emphasized, digital solutions are crucial emerging tools for water utilities to optimize their processes, enhance operational efficiency, and tackle the persistent and growing challenges in the water industry (e.g., Sarni et al., 2019). This trend was also evident in the case study results, where diverse advanced analytical approaches for water networks dominated both in the quantity of signals and the extent of emergence.

This research contributes to the existing literature by developing a framework for quantitative foresight and the identification of emerging trends using large language models. This methodology offers a distinct advantage in terms of scalability, facilitating large-scale studies. Additionally, it is highly adaptable to a variety of data sources, encompassing both structured and unstructured temporal text corpora, which facilitate studying of weak signals from various perspectives. This contributes to the call by Ahlqvist and Uotila (2020) that promote development of analytical steps based on multiple viewpoints and framings, as they emphasize that future objects are complex and cannot be comprehensively understood from a single perspective. It is crucial to recognize that the quality of these data sources can influence the context of the results. For example, using technological data (such as patent data) in the context of water networks primarily emphasizes technological aspects, while scientific publications may also address social topics, including environmental and social sustainability. Moreover, the use of large language models has enhanced the proposed methodology by contextualising signals with greater granularity. While the contextualised signals may create a higher level of complexity in analysing the signals, clustering similar groups of signals enabled us to reserve the important signals. Furthermore, this methodology reduces the reliance on field experts in data examination, making it not only applicable across multiple contexts but also repeatable over various timeframes. This feature enables the observation of changes in emerging trends over time. Also, pertaining to quantitative analysis, we have refined existing indicators for detecting high-growth signals in terms of intensity and diffusion. These new indicators are better tuned to emerging signals, thereby enhancing the overall effectiveness of the methodology.

Beyond the theoretical contributions, this research provides valuable insights for firm managers, researchers, and policymakers. Applying this methodology to various datasets aids in understanding upcoming changes in the environment, enabling stakeholders to prepare for necessary adaptations. In line with Bzhalava et al. (2024), this study provides a methodology where, by integrating weak signals from multiple perspectives, firms can identify entrepreneurial and business opportunities related to weak signals and explore potential products and services. Additionally, this approach allows firms to study areas where business discontinuities might emerge, driven by weak signals. Data-oriented decision making and analysis are crucial for future planning because the exponential increase in data and accelerating environmental complexity render traditional linear decision models obsolete. As the complexity of situations grows, relying solely on outdated models can lead to ineffective and shortsighted decisions. Therefore, integrating comprehensive, cross-disciplinary future studies that embrace the complexity of modern environments is essential (Pratt et al., 2023). By leveraging technological advancements and data-aware methodologies, organizations can better anticipate and navigate future challenges, making informed decisions that are crucial for long-term success. Foresight enhances strategic decision-making by enabling organizations to anticipate and prepare for future changes, engage stakeholders in creating shared visions, explore multiple future scenarios, foster innovation, and continuously adapt to new information and conditions. This integrated approach ensures that strategic decisions are resilient, flexible, and aligned with long-term objectives (Buehring & Bishop, 2020; Vecchiato, 2012).

This research has encountered certain limitations. When choosing the threshold of textual similarity for clustering similar terms into signals, there is a trade-off between precision and duplication in signal representation. Though our choice of threshold, we aimed to optimize this trade-off. However, some signals encompass a wider range of terms than others and representational accuracy may not be consistent across the signals. Additionally, relatively minor changes in the growth rate of quantitative constructs could result in the loss of signals in the final analysis. For future research, there is an opportunity to develop novel methods for signal analysis that extend beyond the measures of intensity or diffusion. Researchers may also consider proposing alternative assessment methodologies that do not solely rely on the growth rate of indicators. Furthermore, future research could investigate the duration signals take to transition from emergence to maturity. This analysis would be grounded in historical data and current well-established trends.

Acknowledgements This study is a part of the SWIM—Smart Water Management project, which was financed by Business Finland through Grant Number 43150/31/2020. This study has also received financial support from another Business Finland funded project, STRA4- Strategic Foresight for Sustainable Innovation and Growth, through Grant Number Grant Number 6863/31/2021.

Credit Author Statements **Sajad Ashouri:** Conceptualization, Methodology, Formal Analysis, Data curation, Writing – original draft; **Carolyn Cole:** Conceptualization, Methodology, Formal Analysis, Data curation, Writing – original draft; **Mona Arnold:** Writing – original draft, **Arash Hajikhani:** Writing – original draft, **Antti-Jussi Tahvanainen:** Writing – original draft, **Arho Suominen:** Revision.

Declaration of Interest None.

References

Ahlqvist, T., & Uotila, T. (2020). Contextualising weak signals: Towards a relational theory of futures knowledge. *Futures, 119*, 102543. https://doi.org/10.1016/j.futures.2020.102543.

Ahuja, G., Coff, R. W., & Lee, P.M. (2005). Managerial foresight and attempted rent appropriation: Insider trading on knowledge of imminent breakthroughs. *Strategic Management Journal, 26*, 791 – 808. https://www.jstor.org/stable/20142269.

Ansoff, H. I. (1975). Managing strategic surprise by response to weak signals. *California Management Review, 18*(2), 21–33. https://doi.org/10.2307/41164635

Ansoff, H. I. (1980). *Strategic Issue Management. Strategic Management Journal, 1*(2), 131–148. https://doi.org/10.1002/smj.4250010204

Ansoff, H. I. (1985). Conceptual underpinnings of systematic strategic management. *European Journal of Operational Research, 19*(1), 2–19. https://doi.org/10.1016/0377-2217(85)90303-0

Ansoff, H. I., Kipley, D., Lewis, A. O., Helm-Stevens, R., & Ansoff, R. (2018). *Implanting strategic management*. Springer.

Baskarada, S., Shrimpton, D., & Ng, S. (2016). Learning through foresight. *Foresight, 18*(4), 414–433.

Bates, M., Grieger, K., Trump, B., Keisler, J., Plourde, K., & Linkov, I. (2015). Emerging technologies for environmental remediation: Integrating data and judgment. *Environmental Science & Technology, 50*(1), 349–358. https://doi.org/10.1021/acs.est.5b03005

Benoit, K., Watanabe, K., Wang, H., Nulty, P., Obeng, A., Müller, S., & Matsuo, A. (2018). Quanteda: an r package for the quantitative analysis of textual data. *The Journal of Open Source Software, 3*(30), 774. https://doi.org/10.21105/joss.00774.

Bibri, S. (2018). Backcasting in futures studies: a synthesized scholarly and planning approach to strategic smart sustainable city development. *European Journal of Futures Research, 6*(1). https://doi.org/10.1186/s40309-018-0142-z.
Bradley, E., Curry, L., & Devers, K. (2007). Qualitative data analysis for health services research: Developing taxonomy, themes, and theory. *Health Services Research, 42*(4), 1758–1772. https://doi.org/10.1111/j.1475-6773.2006.00684.x
Buehring, J., & Bishop, P. C. (2020). Foresight and design: new support for strategic decision making. *She Ji, 6*(3), 408–432. https://doi.org/10.1016/j.sheji.2020.07.002
Buder, F., (2021). The value of foresight in a VUCA world: Results from a survey of organizational foresight capacity. Nuremberg Institute for Market Decisions. https://www.nim.org/en/research/research-reports/research-report-value-foresight-vuca-world.
Bzhalava, L., Kaivo-oja, J., & Hassan, S. S. (2024). Digital business foresight: Keyword-based analysis and CorEx topic modeling. *Futures, 155*, 103303. https://doi.org/10.1016/j.futures.2023.103303.
Calof, J., Meissner, D., & Razheva, A. (2018). Overcoming open innovation challenges: A contribution from foresight and foresight networks. *Technology Analysis & Strategic Management, 30*(6), 718–733. https://doi.org/10.1080/09537325.2017.1351609
Cory, K. (2023). Launching the strategic plan: A critical but missing piece of strategic management education and practice. *Strategy and Leadership, 51*(3), 21–26. https://doi.org/10.1108/sl-11-2022-0112
Daim, T., Monalisa, M., Dash, P., & Brown, N. (2007). Time lag assessment between research funding and output in emerging technologies. *Foresight, 9*(4), 33–44. https://doi.org/10.1108/14636680710773812
Day, G. S. & Schoemaker, P., (2004). Peripheral vision: sensing and acting on weak signals. *Long Range Planning, 2*(37), pp.117–121. https://doi.org/10.1016%2Fj.lrp.2004.01.003.
Devlin, J., Chang, M. -W., Lee, K., & Toutanova, K. (2018). Bert: Pre-training of deep bidirectional transformers for language understanding. *ArXiv Preprint* ArXiv:1810.04805.
Dwivedi, Y. K., Kshetri, N., Hughes, L., Slade, E. L., Jeyaraj, A., Kar, A. K., Baabdullah, A. M., Koohang, A., Raghavan, V., Ahuja, M., Albanna, H., Albashrawi, M.A., Al-Busaidi, A. S., Balakrishnan, J., Barlette, Y., Basu, S., Bose, I., Brooks, L., Buhalis, D., Carter, L., Chowdhury, S., Crick, T., Cunningham, S. W., Davies, G. H., Davison, R. M., Dé, R., Dennehy, D., Duan, Y., Dubey, R., Dwivedi, R., Edwards, J. S., Flavián, C., Gauld, R., Grover, V., Hu, M. -C., Janssen, M., Jones, P., Junglas, I., Khorana, S., Kraus, S., Larsen, K. R., Latreille, P., Laumer, S., Malik, F. T., Mardani, A., Mariani, M., Mithas, S., Mogaji, E., Nord, J. H., O'Connor, S., Okumus, F., Pagani, M., Pandey, N., Papagiannidis, S., Pappas, I. O., Pathak, N., Pries-Heje, J., Raman, R., Rana, N. P., Rehm, S. -V., Ribeiro-Navarrete, S., Richter, A., Rowe, F., Sarker, S., Stahl, B. C., Tiwari, M. K., van der Aalst, W., Venkatesh, V., Viglia, G., Wade, M., Walton, P., Wirtz, J., Wright, R. (2023). "So what if ChatGPT wrote it?" Multidisciplinary perspectives on opportunities, challenges and implications of generative conversational AI for research, practice and policy. *International Journal of Information Management, 71*, 102642. ISSN 0268–4012. https://doi.org/10.1016/j.ijinfomgt.2023.102642.
Ebadi, A., Auger, A., & Gauthier, Y. (2022). Detecting emerging technologies and their evolution using deep learning and weak signal analysis. *Journal of Informetrics, 16*(4), 101344. https://doi.org/10.1016/j.joi.2022.101344.
Frost & Sullivan (2022). *Global Water and Wastewater Treatment, Outlook* (2022)
Flyverbom, M., & Garsten, C. (2021). Anticipation and organization: Seeing, knowing and governing futures. *Organization Theory, 2*(3), 26317877211020324.
Grant, J., & Wunder, T. (2021). Strategic transformation to sustilience: Learning from COVID-19. *Journal of Strategy and Management, 14*(3), 331–351.
Guo, Y., Ma, T., Porter, A. L., & Huang, L. (2012). Text mining of information resources to inform forecasting innovation pathways. *Technology Analysis & Strategic Management, 24*(8), 843–861. https://doi.org/10.1080/09537325.2012.715491

Ha, T., Yang, H., & Hong, S. (2023). Automated weak signal detection and prediction using keyword network clustering and graph convolutional network. *Futures, 152*, 103202. https://doi.org/10.1016/j.futures.2023.103202.

Hamel, G., & Prahalad, C. K., (1994). Competing for the future: a breakthrough strategy for seizing control of your industry and dominating the markets of tomorrow. *Harvard Business School Press.*

Hiltunen, Elina. (2001). Weak signals in organizational futures learning. *Helsinki School of Economics*, 2010.

Hiltunen, E. (2008). Good sources of weak signals: A global study of where futurists look for weak signals. *Journal of Futures Studies, 12*(4), 21–44.

Idrica (2023) Water technology trends 2023 transforming utilities through innovation. 27 p. https://www.idrica.com/resources/.

Karlsen, J. E. (2014). Design and application for a replicable foresight methodology bridging quantitative and qualitative expert data. *Eur J Futures Res, 2*, 40. https://doi.org/10.1007/s40309-014-0040-y

Kaivo-oja, J. R. L., & Lauraeus, I. T. (2018). The VUCA approach as a solution concept to corporate foresight challenges and global technological disruption. *Foresight, 20*(1), 27–49. https://doi.org/10.1108/FS-06-2017-0022

Kim, K., Han, Y., Lee, S., Cho, S., & Lee, C. (2019). Text mining for patent analysis to forecast emerging technologies in wireless power transfer. *Sustainability, 11*(22), 6240. https://doi.org/10.3390/su11226240

LeCun, Y., Bengio, Y., & Hinton, G. (2015). Deep learning. *Nature, 521*(7553), 436–444. https://doi.org/10.1038/nature14539

Lee, H. J., Lee, S., & Yoon, B. (2011). Technology clustering based on evolutionary patterns: The case of information and communications technologies. *Technological Forecasting and Social Change, 78*(6), 953–967. https://doi.org/10.1016/j.techfore.2011.02.002

Li, M., Porter, A. L., Suominen, A., Burmaoglu, S., & Carley, S. (2021). An exploratory perspective to measure the emergence degree for a specific technology based on the philosophy of swarm intelligence. *Technological Forecasting and Social Change, 166*, Article 120621. https://doi.org/10.1016/j.techfore.2021.120621

Liu, L., Chiang, C., & Chen, S. (2023). Patent trend analysis: extended reality (xr) and future virtual adventure., 27–45. https://doi.org/10.47260/amae/1322.

Liu, Y., Zhou, Y., & Ling, L. (2016). Application of big data analysis method in technology foresight for strategic emerging industries. *Chinese Journal of Engineering Science, 18*(4), 121. https://doi.org/10.15302/j-sscae-2016.04.018.

Mao, C., Koide, R., Brem, A., & Akenji, L. (2020). Technology foresight for social good: Social implications of technological innovation by 2050 from a Global Expert Survey. *Technological Forecasting and Social Change, 153*, Article 119914. https://doi.org/10.1016/j.techfore.2020.119914

Marinkovic, M., Al-Tabbaa, O., Khan, Z., & Wu, J. (2022). Corporate foresight: A systematic literature review and future research trajectories. *Journal of Business Research, 144*, 289–311. https://doi.org/10.1016/j.jbusres.2022.01.097

Martinsons, M., Everett, J., & Chan, K. (2001). Mapping the scholarly development of strategic management. *Land Contamination & Reclamation, 27*(2), 101–110. https://doi.org/10.1177/0165551014233590

Mendonça, S., e Cunha, M. P., Kaivo-oja, J. and Ruff, F., (2004). Wild cards, weak signals and organisational improvisation. *Futures, 36*(2), pp.201–218. https://doi.org/10.1016/S0016-3287(03)00148-4.

Moor, J. H. (2005). Why we need better ethics for emerging technologies. *Ethics and Information Technology, 7*(3), 111–119. https://doi.org/10.1007/s10676-006-0008-0

Moorman, C., & Miner, A. S. (1998). The convergence of planning and execution: Improvisation in new product development. *Journal of Marketing, 62*(3), 1–20. https://doi.org/10.1177/002224299806200301

Mühlroth, C., & Grottke, M. (2020). Artificial intelligence in innovation: how to spot emerging trends and technologies. *IEEE Transactions on Engineering Management, 69*, 493–510. https://doi.org/10.1109/TEM.2020.2989214
Nakamoto, S. (2008). Bitcoin: A peer-to-peer electronic cash system.
Popper, R. (2008). How are foresight methods selected?. *Foresight, 10*(6), pp.62–89.
Pratt, L., Bisson, C., & Warin, T. (2023). Bringing advanced technology to strategic decision-making: The Decision Intelligence/Data Science (DI/DS) Integration framework. *Futures, 152*, Article 103217.
Reeves, M. & Deimler, M. (2011). Adaptability: the new competitive advantage. *Harvard Business Review* 2011/07.
Reimers, N., & Gurevych, I. (2019). Sentence-bert: Sentence embeddings using siamese bert-networks. *ArXiv Preprint* ArXiv:1908.10084.
Rohrbeck, R., Battistella, C., & Huizingh, E. (2015). Corporate foresight: An emerging field with a rich tradition. *Technological Forecasting and Social Change, 101*, 1–9. https://doi.org/10.1016/j.techfore.2015.11.002
Rohrbeck, R., & Kum, M. E. (2018). Corporate foresight and its impact on firm performance: A longitudinal analysis. *Technological Forecasting and Social Change, 129*, 105–116. https://doi.org/10.1016/j.techfore.2017.12.013
Rohrbeck, R., & Schwarz, J. O. (2013). The value contribution of strategic foresight: Insights from an empirical study of large European companies. *Technological Forecasting & Social Change, 80*, 1593–1606. https://doi.org/10.1016/j.techfore.2013.01.004
Rotolo, D., Hicks, D., & Martin, B. R. (2015). What is an emerging technology? *Research Policy, 44*(10), 1827–1843.
Sahin, U., Karikó, K., & Türeci, Ö. (2014). MRNA-based therapeutics—developing a new class of drugs. *Nature Reviews Drug Discovery, 13*(10), 759–780. https://doi.org/10.1038/nrd4278
Sarni, W., White, C., Webb, R., Cross, K. & Glotzbach, R. (2019). Industry leaders chart the transformation journey. *International Water Association (IWA) and Xylem White Paper*.
Sutherland, W. J., & Woodroof, H. J. (2009). The need for environmental horizon scanning. *Trends in Ecology & Evolution, 24*(10), 523–527. https://doi.org/10.1016/j.tree.2009.04.008.
von Ditfurth, H., Weisbord, E., Danielsen, T., Zutari, L.F.J., Hafemann, A.C., Hima, J. and Oraeki, T.C., 2021. Digital Water: An Overview of the Future of Digital Water from a YWP Perspective. Semke, L.-M. & Tiberius, V. (2020). Corporate foresight and dynamic capabilities: An exploratory study. *Forecasting, 2*(2), 180–193.
Schoemaker, P. J., & Day, G. (2021). Preparing organizations for greater turbulence. *California Management Review, 63*(4), 66–88.
Schwarz, J. O., Rohrbeck, R., & Wach, B. (2020). Corporate foresight as a microfoundation of dynamic capabilities. *Futures & Foresight Science, 2*(2). https://doi.org/10.1002/ffo2.28.
Ting, P., Shan, X., & Jiang, G. (2012). *Research on Risk Assessment of Emerging Technology Industrialization Based on Gray Clustering*. https://doi.org/10.1109/ismot.2012.6679471
Tsoukas, H., & Shepherd, J. (2004). Organisations and the future: From forecasting to foresight. *Management Today, 20*(7), 18–23.
Vecchiato, R. (2012). Environmental uncertainty, foresight and strategic decision making: An integrated study. *Technological Forecasting and Social Change, 79*(3), 436–447. https://doi.org/10.1016/j.techfore.2011.07.010
Uhlenbrook, S., & Connor, R. (2019) The United Nations world water development report 2019: leaving no one behind.
Vecchiato, R. (2015). Creating value through foresight: First mover advantages and strategic agility. *Technological Forecasting & Social Change, 101*, 25–36. https://doi.org/10.1016/j.techfore.2014.08.016
Wang, Y., & Zhang, C. (2020). Using the full-text content of academic articles to identify and evaluate algorithm entities in the domain of natural language processing. *Journal of Informetrics, 14*(4), Article 101091.

Wenzel, M. (2022). Taking the future more seriously: From corporate foresight to "future-making." *Academy of Management Perspectives, 36*(2), 845–850.

Wiebe, K., Zurek, M., Lord, S., Brzezina, N., Gabrielyan, G., Libertini, J., Loch, A., Thapa-Parajuli, R., Vervoort, J., & Westhoek, H. (2018). Scenario development and foresight analysis: Exploring options to inform choices. *Annual Review of Environment and Resources, 43*, 545–570. https://doi.org/10.1146/annurev-environ-102017-030109

Wu, Y., Lin, Z., Zhao, Y., Qin, B., & Zhu, L. (2021). A text-centered shared-private framework via cross-modal prediction for multimodal sentiment analysis. https://doi.org/10.18653/v1/2021.findings-acl.417

Xu, S., Hao, L., An, X., Yang, G., & Wang, F. (2019). Emerging research topics detection with multiple machine learning models. *Journal of Informetrics, 13*(4), Article 100983.

Yoon, J. (2012). Detecting weak signals for long-term business opportunities using text mining of Web news. *Expert Systems with Applications, 39*(16), 12543–12550. https://doi.org/10.1016/j.eswa.2012.04.059

Creation of a Unit of the ICAT's Surgery Simulation Laboratory at the Hospital General De México Based on Scientometric and Bibliometric Analysis and a Technology Roadmap Construction

L. R. Vega-González, M. A. Padilla-Castañeda, N. Castillo-Camarena, J. S. Pérez-Lomelí, and C. F. Domínguez-Velasco

1 Introduction

Porras-Hernández (2016) defines surgery as a manual work that involves the mechanical manipulation of the anatomical structures of a human being for medical purposes. It is the branch of medicine that prevents, cures, or rehabilitates diseases in patients by cutting, separating, repairing, or replacing tissues or organs with instruments, usually under anesthesia. It is an eminently action-focused practical discipline. As a collection of procedures used to restore or maintain the health of a human being, surgery is a technology.

The author points out that, even in the best centers in the world, they have a risk of death of 1% and injury of 8%. The more effective interventions in the practice

L. R. Vega-González (✉) · N. Castillo-Camarena
Secretaría de Vinculación y Gestión Tecnológica, Instituto de Ciencias Aplicadas y Tecnología, Universidad Nacional Autónoma de México, Mexico, México
e-mail: roberto.vega@icat.unam.mx

N. Castillo-Camarena
e-mail: nadia.castillo@icat.unam.mx

M. A. Padilla-Castañeda · J. S. Pérez-Lomelí · C. F. Domínguez-Velasco
Grupo de Bioinstrumentación, Instituto de Ciencias Aplicadas y Tecnología, Universidad Nacional Autónoma de México, Mexico, México
e-mail: miguel.padilla@icat.unam.mx

J. S. Pérez-Lomelí
e-mail: salvador.perez@icat.unam.mx

C. F. Domínguez-Velasco
e-mail: cesar.dominguez@icat.unam.mx

H. da Hora et al. (eds.), *Technology Mining*,
https://doi.org/10.1007/978-3-032-10849-4_6

of safe or excellent surgery are based on the teaching and learning of procedures, attitudes, and values. For example, the surgeon profile established by the Mexican Council of Pediatric Surgery (Consejo Mexicano de Cirugía Pediátrica, in Spanish) for Mexican surgeons is organized into three levels of performance: (1) Knowledge and skills: what a good surgeon should know and be able to do, (2) Attitudes: how a good surgeon approaches the practice of Surgery and (3) Personal qualities: how a good surgeon behaves as a medical professional.

The first level depends on the theoretical and *practical* preparation of the surgeon. This knowledge and abilities are obtained at universities, medical schools, clinics, and hospitals. Worldwide, teaching and training in surgery has been based for almost two centuries on the master-apprentice model. Recently, the legislation and the availability of patients and hospitals to practice, have reduced the number of training hours for young surgeons to become specialists. The number of flight hours has been reduced by cutting the training time of surgery students. However, acquiring safe operative techniques is only one of the many attributes required to be a good surgeon. Surgeons also need a core knowledge base, clinical decision-making ability, communication skills, drive, ambition, and the ability to think and work under stress (Spencer, 1978, cited by Torkington et al., 2000).

Undoubtedly, there is a potential difficulty in obtaining enough hands-on operative experience. A good general surgeon must perform many cases across the whole range of general surgery, and a specialist surgeon must require extra intensive hands-on preparation. For Torkington et al. (op cit.), the main questions to answer are: How did we manage to train new competent specialists without exposing the patients to possible consequences of their inexperience? How do we weigh the need for acquisition skills and practical knowledge of residents with the security of any procedure when intervening on a human being?

Fortunately, the answer to those questions came from the spectacular advances in information technology information and telecommunications that occurred in the last decades, which have changed the way it develops virtually every form of human activity, including the surgical medical profession. Today, several simulation techniques are available, including artificial tissues, animal models, and virtual reality computer simulation. Nowadays, the possibility of training surgical skills outside the operating theatre is a reality.

In Mexico, the use of simulators for medicine teaching applications began in the eighties of the last century, with high-fidelity simulators in some universities. In 2003, the National Institute of Nutrition "Salvador Zubirán" created the Center for the Development of Medical Skills (CEDDEM). Almost immediately, in 2005, the UNAM's Medicine Faculty created the Center for Teaching and Certification of Medical Aptitudes (CECAM). The Center for the Improvement of Medical Abilities and Skills of the Directorate of Health Education at the Hospital General de México Dr. Eduardo Liceaga (HGMEL) was also created over more than two decades ago.

Manrique-Gutiérrez et al. (2021) found that after the COVID-19 pandemic, there were needed changes and new strategies in medical education. The social distancing that drastically altered the way of carrying out daily activities caused the need to go from face-to-face teaching to a completely virtual format. This motivated the emergence of various platforms, programs, and digital technologies in the medical curricular plan, which included 3D programs and virtual simulators to put clinical skills into practice.

In addition, augmented reality programs with 3D models were needed in subjects such as Anatomy to replace cadaveric dissections, while virtual microscopes were used in pathology. All this was possible with the emergence of electronic accessories such as compact cameras and camcorder glasses that guide the recording of surgeries in real time that students can observe without participating in person.

Similar situations were experienced in other Latin American countries. Mastroianni et al. (2020) also reported changes made in the Surgery Residency Program of an Argentine University Hospital aimed at providing the same educational quality in a safe environment during the COVID-19 pandemic.

Virtual teaching was an essential tool for medical learning during the health contingency caused by Sars-Cov-2. However, we can say that digital medical education worldwide is here to stay. In that regard, *the purpose of this work* is to present the case of the Creation of a Unit of the Surgery Simulation Laboratory of the Applied Science and Technology Institute (ICAT) of the National Autonomous University of México at the HGMEL General Hospital of México (HGMEL). To develop the plan for the creation of the Unit, there were some stages. First, it was realized scientometric, bibliometric, and market data analysis; after, having in mind some of these results, a Technology Roadmap (TRM) was built. The results clearly pointed out that the future of surgery teaching will be digital simulation and advanced computational techniques such as virtual and augmented reality.

2 Framework

Naturally, insufficient hands-on preparation has led to a host of problems. Berner and Ewertz (2018) state that medical errors of preventable causes are estimated to result in more than 400,000 deaths per year. Annually, in the United States, these are the third cause of death in this country, while non-fatal iatrogenesis generates disability in more than 3.5 million patients annually. On the other hand, the judicialization of medical practice leads to medical errors being practically unacceptable. The phenomenon has impacted the training of specialists, especially in those branches involved in invasive surgical procedures. Therefore, patient safety is essential in any medical intervention.

Berner and Ewertz, (op cit) claim that several factors are identified as threats for medical education in the current healthcare scenario. Increased litigations, working-time directives, patient safety issues, and a higher social demand for specialists have forced medical educators to look for new alternatives to traditional teaching. In this scenario, simulation provides a recreation of a clinical situation.

Scientometric and Bibliometric Information on Surgery Training

We used the Scopus database as the source of information for the bibliometric analysis and VOSviewer software to process the information. The analysis helps identify the stakeholders, the relevance of the topic, and the possible lines of research to explore and develop.

Surgery training is a topic with a growing trend in its publication in the last 20 years. The countries with higher publication numbers are shown in Table 1 and Fig. 1 shows joint publication between countries.

Table 1 Countries with higher production between 2010–2021

	Country	Nr. documents		Country	Nr. documents
1	USA	792	11	Spain	39
2	United Kingdom	162	12	India	32
3	Canada	127	13	Switzerland	29
4	China	109	14	Netherlands	26
5	Germany	85	15	Brazil	21
6	Turkey	64	16	Singapore	19
7	Japan	60	17	Austria	19
8	France	51	18	Saudi Arabia	19
9	Australia	46	19	Iran	17
10	Italy	43	20	Belgium	17

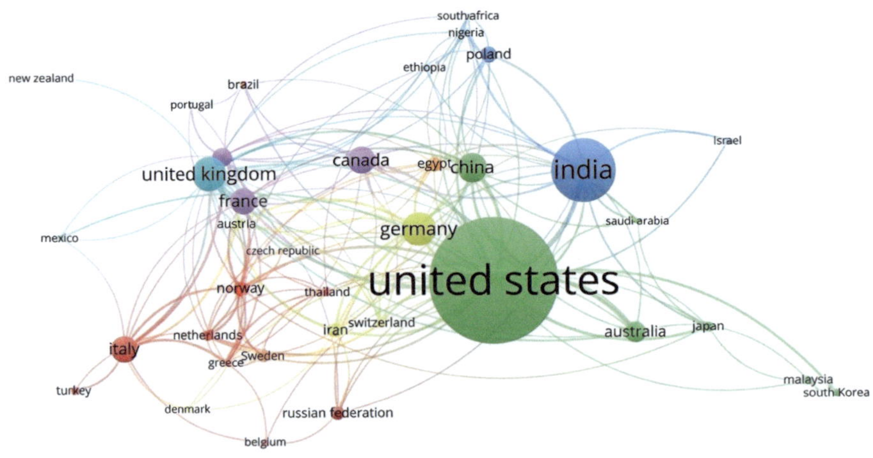

Fig. 1 Joint publication between countries on surgery training

The United States of America (USA) is the leading country publishing on the topic of interest. The USA has a strong collaboration relationship with many countries, mainly Canada, the United Kingdom, Germany, China, and France. Other countries publishing with the USA are Kenya, Italy, Belgium, Turkey, Japan, and Brazil. The size of the circles indicates the number of documents published. In contrast, the relationships are indicated by the thicker lines when more documents are published together.

Medicine is the main area of study in which the keyword surgery training is published, followed by Computer Science with seven times fewer publications, then Engineering, and finally, Social Sciences.

The 20 institutions with the largest number of documents published in the period 2010–2021, the subject are: (1) the University of Toronto (41), (2) the University of Pennsylvania (29), (3) the University of Michigan Ann Harbor (28), (4) Stanford University (27), (5) Northwestern University (25), (6) Massachusetts General Hospital (25), (7) Harvard Medical School (24), (8) University of Washington (24), (9) Baylor College of Medicine (22), (10) John Hopkins University (21), (11) Mayo Clinic (21), (12) Oregon Health and Science University (21), (13) The University of British Columbia (20), (14) University of California San Francisco (19), (15) The University of North Carolina at Chapel Hill (18), (16) Brigham and Women's Hospital (17), (17) Cleveland Clinic Foundation (17), (18) Université McGill (16), (19) Northwestern University Feinberg School of Medicine (16), (20) the University of Texas MD Anderson Cancer Center (16). The journals with 15 or more documents are shown in Table 2.

The principal authors with the largest number of papers are presented in Table 3. Authors from the USA are the greater production in the field of knowledge.

3 Keyword Co-Occurrence

There were 7,324 keywords we used for the co-occurrence analysis of all the keywords and a full counting method. A minimum of 5 occurrences of a keyword was used for the analysis. 863 keywords meet the threshold. The map shows six clusters in which the keywords are grouped. The green and blue have to do with the characteristics of the patients or studies. The light blue keywords refer to procedures and studies. The green keywords refer to the medical education and curricula. The yellow group of keywords refers to external elements of the field of study. In general, all keywords are recently used in the field of study.

Surgery training and virtual reality are related to education, surgery, laparoscopy, simulation, and procedures, among others.

The keyword map is shown in Fig. 2.

USA institutions are the top producers of documents in the field of surgery training. The institution with the largest number of documents published in the period is the University of Toronto, Canada (see Table 4).

Table 2 Main information sources

	Publications	No. Documents
1	Journal of Surgical Education	105
2	Plastic and Reconstructive Surgery	55
3	Surgical Endoscopy	37
4	American Journal of Surgery	34
5	Journal of Vascular Surgery	32
6	Annals of Thoracic Surgery	31
7	Journal of Pediatric Surgery	31
8	Annals of Plastic Surgery	30
9	Journal of Thoracic and Cardiovascular Surgery	29
10	Surgery United States	28
11	Journal of Surgical Research	19
12	Plastic and Reconstructive Surgery Global Open	18
13	Journal of Plastic Reconstructive and Aesthetic Surgery	17
14	Journal of Robotic Surgery	17
15	Lecture Notes in Computer Science Including Subseries Lecture Notes in Artificial Intelligence and Lecture Notes in Bioinformatics	17
16	Annals of Vascular Surgery	15
17	Journal of Oral and Maxillofacial Surgery	15
18	Aesthetic Surgery Journal	14
19	World Journal of Surgery	13
20	Annals of Surgery	12
21	ANZ Journal of Surgery	11
22	American Surgeon	10
23	International Journal of Medical Robotics and Computer Assisted Surgery	10
24	JAMA Surgery	10

Surgery training, limited to Mexico, yielded 67 related institutions. Institutions with two or more results are shown in Table 5. In the list there are Mexican institutions with collaborations with authors of other countries institutions.

As simulation is a relevant theme in surgery training, we present next the results of a bibliometric analysis regarding surgery simulation.

Bibliometric Information on Surgery Simulation

In Fig. 3, we have joint publications on surgery simulation between countries to get around the theme of using digital tools for surgery simulation. The USA and China

Table 3 Most prominent authors

	Author	Institution	Country	No. documents
1	Guo, S	University of Toronto	Canada	16
2	Gao, B	University of Pennsylvania	USA	11
3	Daniels, A.H	University of Michigan, Ann Arbor	USA	10
4	Ivaschenko, A	Stanford University	USA	10
5	Janis, J.E	Northwestern University	USA	8
6	Kolsanov, A	Massachusetts General Hospital	USA	8
7	Lou, X	Harvard Medical School	USA	8
8	Motaharifar, M	University of Washington	U.S.A	8
9	Stephens, E.H	Baylor College of Medicine	USA	8
10	Tchantchaleishvili, V	Johns Hopkins University	USA	8
11	Chung, K.C	Mayo Clinic	USA	7
12	Kuzmin, A	Oregon Health and Science University	USA	7
13	Taghirad, H.D	The University of British Columbia	U.S.A	7
14	Xiao, N	University of California, San Francisco	U.S.A	7

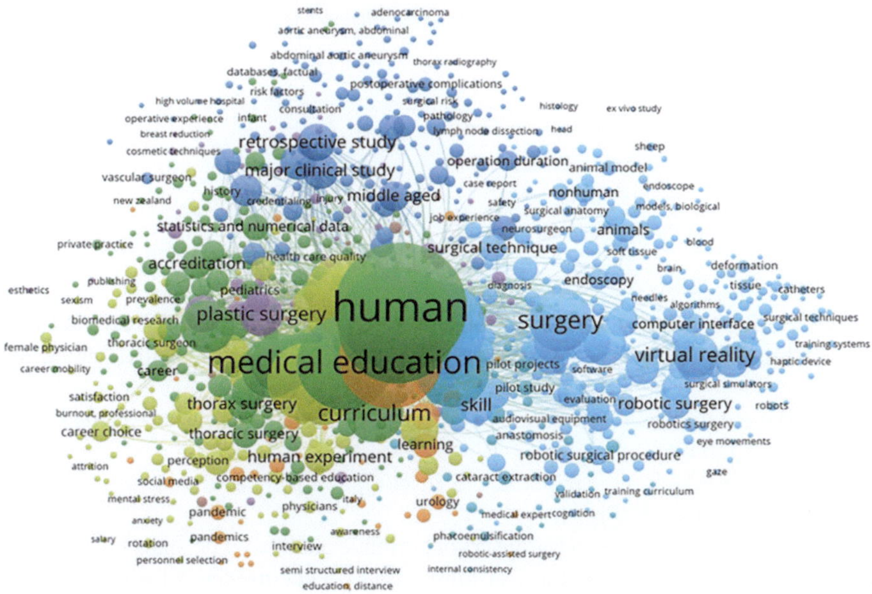

Fig. 2 Keywords related to surgery training

Table 4 Most relevant institutions on surgery training

	Institution	Country	No. documents
1	University of Toronto	Canada	16
2	University of Pennsylvania	USA	11
3	University of Michigan, Ann Arbor	USA	10
4	Stanford University	USA	10
5	Northwestern University	USA	8
6	Massachusetts General Hospital	USA	8
7	Harvard Medical School	USA	8
8	University of Washington	USA	8
9	Baylor College of Medicine	USA	8
10	Johns Hopkins University	USA	8
11	Mayo Clinic	USA	7
12	Oregon Health and Science University	USA	7
13	The University of British Columbia	USA	7
14	University of California, San Francisco	USA	7

Table 5 Most active Mexican institutions in surgery training

	Institution	No. documents
1	National autonomous university of Mexico	6
2	Center for research and advanced studies of the national polytechnic institute	4
3	General hospital of Mexico	4
4	Autonomous university of Nuevo León	3
5	University hospital Dr. José Eleuterio González	3
6	Children's hospital of Mexico Federico Gómez	2
7	Mexican social security institute	2
8	National nutrition institute Salvador Zubirán	2

are the leading countries where the topic is being studied and published, and their relationship is strong. Other important players are Germany, Canada, Japan, and France.

The 20 institutions with the largest number of documents published on the subject are shown in Table 6.

The keywords that appear published in conjunction with the keyword surgery simulation are shown in Fig. 4.

5,125 keywords were used for the co-occurrence analysis. The analysis was made with a fractional counting method. A minimum of 5 occurrences of a keyword was used. 438 keywords meet the threshold. The map shows the clusters in which the keywords are group, in this case 5 clusters. The green cluster has keywords related to training and surgical procedures. The yellow cluster contains words related to

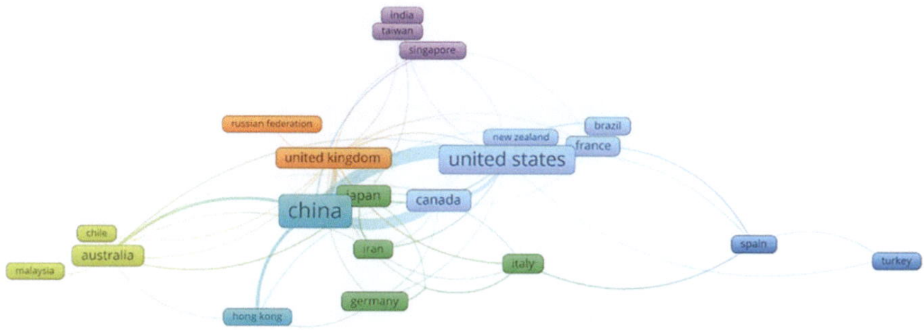

Fig. 3 Joint publication between countries on surgery simulation

Table 6 The 20 institutions with the largest number of documents published on the subject surgery simulation

	Institution	Documents		Institution	Documents
1	Beihang university	21	11	Fujita health university	9
2	Nanchang university	16	12	Sharif university of technology	8
3	Shanghai Jiao Tong university	16	13	Chinese university of Hong Kong	7
4	Old Dominion university	13	14	Southeast university	7
5	Carleton university	13	15	Institute of automation Chinese academy of sciences	7
6	Nagoya university	12	16	Monash university	7
7	Rensselaer polytechnic institute	11	17	Samara state medical university	6
8	Chinese academy of sciences	10	18	Taipei medical university hospital	6
9	Nanjing university of information science & technology	10	19	The university of Western Ontario	6
10	Harbin engineering university	9	20	RMIT university	6

computer simulation. The red cluster has to do with software to develop simulations. The blue cluster, closest to the yellow training cluster, has to do with images and models. Finally, the purple cluster, with fewer keywords, is made up of words related to analysis.

In the case of Mexico, 11 publications were found between 2003 and 2021. For the period 2010–2021, up to now, there are 11 documents. It is observed that the first document in Scopus database in this topic that has to do with Mexico is from 2003.

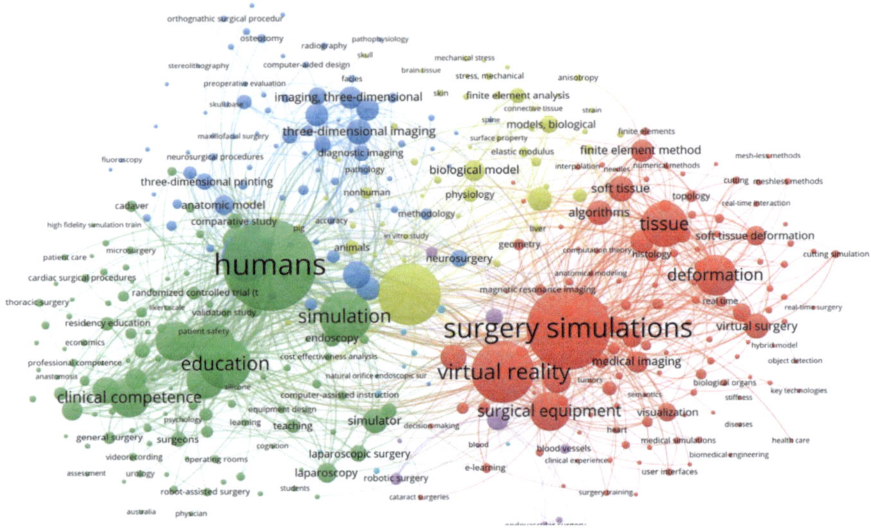

Fig. 4 Keywords related to *surgery simulation.*

The socialization of research in surgery simulation in Mexico has two decades, at least.

The Universidad Nacional Autónoma de México has 9 publications. Likewise, the Hospital General de México, the CINVESTAV Unidad Guadalajara, and the Instituto Nacional de Astrofísica Optica y Electrónica have one publication each.

The bibliometric analysis shows that simulation and surgery training are topics of interest to the academic community. In Mexico, they represent an opportunity for research and development. The main keywords related to *surgery training* obtained in Fig. 2 are medical education, surgery, virtual reality, surgical technique, mayor clinical study, plastic surgery, and skills. The main keywords related to *surgery simulation* of Fig. 4 are virtual reality, simulation, surgery simulations, surgical equipment, education, clinical competence, and virtual surgery. Finally, the keywords related to *surgery simulation* in Mexico of Fig. 5 are virtual reality, surgery simulation, computer simulation, surgery, and tissue.

4 Technology Roadmap a Planning Technique

Technology roadmap (TRM) is a plan formulation tool used since the 1970s with Motorola's initial development of the technique. TRM is a process used to plan and communicate an organization's technology strategy. Albright (2003) considered the TRM a support tool in managing and planning technologies. For Loyarte et al. (2015), TRM is a strategic tool for organizations willing to develop a vision.

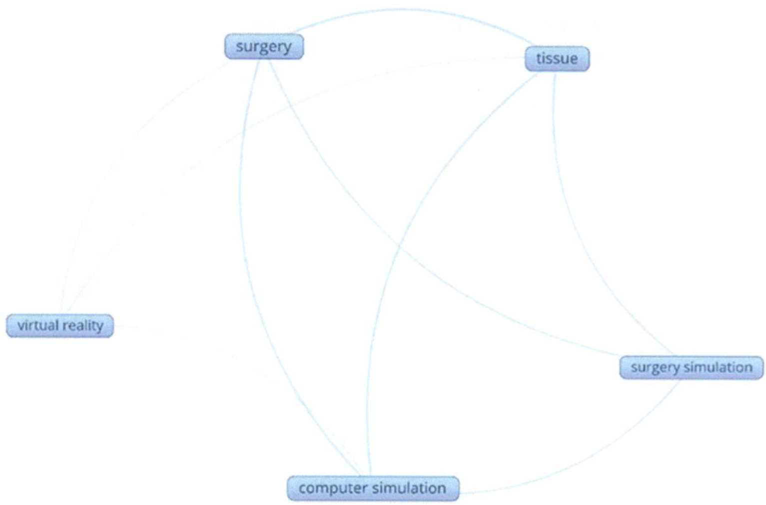

Fig. 5 Keywords related to *surgery simulation* in Mexico

Phaal et al., (2001a, 2001b) indicate that TRM "can be helpful to support and communicate technology strategy and planning, comparing market opportunities with product and technology development" (pg. iii). Groenveld (1997) identifies TRM as a process that contributes to an organization, with the integration of technology and the definition of technology strategy, by visualizing the interaction between products and technologies over time, considering both short and long-term products and technology aspects. TRMs help to identify the resources needed to develop a technology.

Carvalho et al. (2013) point out that there is an opportunity in the literature of TRM regarding situations between TRM and strategic alignment of an organization. Another area that has received little attention in the literature is the interaction of TRM and knowledge management (De Alcantara & Martens, 2019).

Undoubtedly, TRM is a powerful tool for supporting technology management and planning in an organization. It is a flexible process, as shown by Phaal et al. (2001), who identified different types of TRM according to the target purpose: product planning, service/capability planning, *strategic planning*, long-range planning, knowledge asset planning, program planning, process planning, and integration planning. Even more, in a subsequent paper, the authors identified six types of areas in which TRM has been applied: strategic evaluation, business reconfiguration, product-technology planning, process development, research network development, and sector foresight (Probert, Farrukh, & Phaal, 2003).

Medical Simulators Emergence

According to CEPAL (2021) (Economic Commission for Latin America and the Caribbean, by its acronym in Spanish), the information society and the advances in telecommunications, the world digital agenda, the integration of advanced digital technologies such as fifth generation (5G) mobile networks, the Internet of Things (IoT), cloud computing, artificial intelligence, data analytics, big data, data science, robotics, augmented reality, among others, will bring to society well-being, productivity efficiency and effectiveness, and will contribute for the sustainability for all economic sectors and areas of human activity. Of course, this is extensive for teaching and training human resources, including medicine teaching, particularly for training highly specialized surgeons.

The evidence provided by direct interviews with heads of medical services in various Mexican health institutions clearly indicates that *simulation for teaching surgery is the trend observed worldwide to prepare surgeons of all specialties.* Sánchez-González et al. (2018) suggest, soon, there will be a worldwide new virtual reality-based environment for surgical training.

The salient features of this new teaching approach are minimally invasive surgery, content creators, and professors empowered to create their own training scenarios with clinical cases of patients, but in safe and patient-free environments. Also, it is expected to reduce the learning curves, train more confident professionals, and expand the accessibility of training to a broader population of medical residents.

In some countries, such as the United States, clinical and surgical simulation training is already included in some resident study programs for various specialties. Within the study programs, residents cover a minimum number of hours of practice using anatomical models and mannequins for basic skills, and simulators for specialized skills, in conditions closer to reality before having access to interventions with real patients. In this way, the teaching processes of these study programs are designed in phases in which residents acquire progressively more complex skills, optimizing available clinical resources and avoiding risks to patients. Moreover, even avoiding risks to themselves, such as excessive exposure to ionizing energy and pathogens.

There are several examples of this type of teaching units in the USA. For example, the Western University of Health Sciences in Pomona, California, has a training center designed for medical and nursing students. It uses virtual reality, robotics, artificial intelligence, and simulation technologies as part of its training platform. Another notable example is the Neurosurgical Simulation Center and Virtual Reality from Stanford University.

In Europe, there are specialized research centers for the development and technological innovation of new surgical techniques and their teaching directly in hospitals. For example, together, the University of Pisa and the Regional Hospital of Cisanello in Pisa, Italy, have installed the ENDOCAS center within the Hospital, a laboratory for the development and research of new techniques and surgical planning and training, as well as intraoperative assistance, based on the use and development of simulators.

In Latin America, Brazil has established medical education programs and technology-based companies for the health sector through University-Hospital synergy to install research and innovation centers to develop new simulation technologies within hospitals.

Digital simulation seems to be the dominant future trend for universities, hospitals, and medical teaching centers in Mexico. In this country, there are university centers for teaching and medical skills at the undergraduate and postgraduate level in the leading universities such as UNAM, the Universidad Autónoma de Nuevo León (UANL, for its acronym in Spanish), and the Anahuac University, among others, in their Faculties of Medicine and research centers. However, there is still a substantial gap in establishing itself as a standard for teaching directly in health institutions. Regarding hospitals and health centers, the current situation is that there is increasingly limited access to biological models and cadavers due to legal, economic, and infrastructure restrictions, as well as a shortage of animal models for bioethical reasons.

Nowadays, there are some simulation centers, such as the National Medical Center "Century XXI" of the Mexican Social Security Institute (IMSS, for its acronym in Spanish), among others. These centers have adopted the use of commercial training systems, primarily through low and medium-fidelity models. However, their high costs, technical limitations, and low availability still make their access difficult in practice. These factors, together with the increased workload due to the high demand of patients, make it necessary to design and implement alternatives that improve residents' teaching processes, reduce the risk for patients, and optimize available resources.

Using TRM to Define a Project to Establish a Surgery Simulation Lab at the Hospital General De México

The evidence provided by the keywords obtained in the Figs. 2, 4 and 5 of the scientometric and bibliometric studies and the technology foresight based on the TRM and direct interviews with heads of medical services in various Mexican health institutions clearly indicates that *simulation for teaching surgery is the trend observed worldwide to prepare surgeons of all specialties.*

Over the course of a couple of years, the working team of ICAT and the HGMEL Teaching Unit discussed the need and convenience of installing the Surgery Simulation Lab in the Hospital's Teaching Unit in order to improve the training procedures of the Hospital's medical residents. Part of the planning was the construction of the TRM since according to Alvarenga dos Santos et al. (2022), TRM aims to answer three questions: where are we going? Where are we now? And how will we get there? Using these questions as a guide, several meetings were held with the counterpart (HGM) to identify at least one of their needs that could be filled with the ICAT's technological capabilities.

Using a basic building TRM technique, a standard type of roadmap was developed, which contains a multilayer time-based chart on which the markets, products, and technology are mapped, showing key links between layers (Phaal, Farrukh, & Probert, 2001a, 2001b).

This planning tool confirmed the need to install a Surgery Simulation Laboratory pursuing the strategic transformation of the Center for the Improvement of Medical Skills (CPHDM) of the General Hospital of Mexico "Dr. Eduardo Liceaga". The TRM obtained for a clinical simulator research and development laboratory is shown in Fig. 6. The dotted line blocks at the base of the TRM express the technological skills developed over the years and which are part of the domain of the ICAT Bioinstrumentation Group. The solid line blocks shown in the central part of the TRM show technologies already developed and which have intellectual protection, property of UNAM. Finally, the upper rectangular box shows the training areas already defined for the HGMEL Teaching Unit.

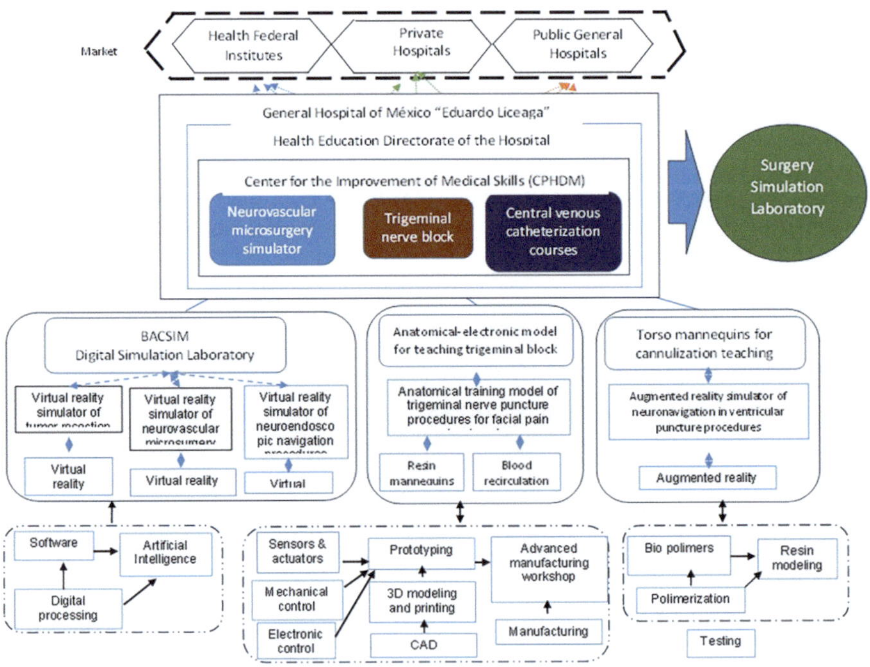

Fig. 6 TRM for a surgery simulation laboratory

5 Case Study: A Plan for the Creation of the Surgery Simulation Laboratory (SSL)

A case study of the creation of a Unit of the digital Surgery Simulation Laboratory (SSL) of the ICAT at the Hospital General de México "Eduardo Liceaga" (HGMEL, by its acronym in Spanish) is presented. The plan to create this digital transdisciplinary university-hospital high-social impact project was driven based on the information and data obtained using scientometric and bibliometric analysis and the technology roadmap presented in the previous section.

Background: *The Development of the ICAT's UIDT at the HGMEL*

The HGMEL was founded on February 5, 1905, by President Porfirio Diaz. It is currently named after its founder, Dr. Eduardo Liceaga, who was its first Medical Director in Spanish as Hospital General de México "Dr. Eduardo Liceaga". HGMEL is a public assistance institution belonging to the Ministry of Health of Mexico. It is part of a network of eleven public general hospitals in Mexico, that provide public health services in the central area of the country, located in Colonia Doctores, a popular surrounding quarter, cataloged as a high-density populated and low-medium income socio-demographic quarter, a few steps from the General Hospital subway station, line 3 of Mexico City (HGM 2022).

The HGMEL has three sections: medical specialties, surgery, diagnostics, and treatment. Its medical personnel give more than 1,000 consultations for different medical specialties daily. Naturally, it is a multidisciplinary organization, considered as Level 3 according to the Mexican Health public system, where doctors of various medical specialties and high-specialties, in collaboration with engineers and technicians for maintenance, biomedical engineers for imaging, X Rays, equipment, and instrumentation support and administrative and juridical personnel among many others. In general, the range of population it serves corresponds, in most cases, to patients from poor strata, many coming from places far from the center of the country. Therefore, the demand for services, workload, and the Hospital's financial requirements are very high, at the limit of its capabilities.

From a bibliometric analysis, the HGMEL publishes scientific reports and articles mainly alone. However, it has joint publications mainly with the National Autonomous University of Mexico, the Mexican Institute of Social Security, and the National Institute of Medical Sciences and Nutrition Salvador Zubirán. The academic relationship with the UNAM comes from the sixties of the past century. Later, by 1971, the Experimental Medicine Unit of UNAM (UEMU) was created and established within the Hospital. In this Unit, medical personnel from HGMEL collaborate with medical doctors from the Faculty of Medicine of the UNAM to develop *scientific* projects, mainly in basic biomedicine and immunology research (Facultad de Medicina, 2022).

It took more than four decades after the creation of the UEMU for the creation of a new entity focused on *technological development*, the Research and Technological Development Unit (UIDT) of the Applied Sciences and Technological Development Center (CCADET, from Spanish) within the General Hospital of Mexico, "Eduardo Liceaga" (HGMEL). The UIDT was developed because of the technological alliance between the UNAM and the Hospital. The alliance could be done thanks to the negotiations and joint work of the authorities of the HGMEL and UNAM. By 2011, a series of projects of interest for the parties were foreseen.

To formalize the UIDT's creation, UNAM, through the ICAT and the HGMEL, signed on February 29, 2012, a Specific Collaboration Agreement to carry out interdisciplinary translational research and medical technology development. Such activities mainly in the development of new methods and technological solutions for better diagnostics and treatment, related to pathologies and diseases mainly that affect the health of the Mexican people and according to the Hospital's interest. The UIDT was inaugurated on September 20, 2013.

6 The UNAM-ICAT's/HGMEL Research and Technological Development Unit (UIDT) Today (ICAT, 2023)

Initially, the main objective of the UIDT was "to develop research and technology on new materials, devices, equipment, software, diagnostics, and therapeutic procedures, as well as support for professional practice and clinical teaching related to diseases that affect the general population's health", (Vega-González, García-Segundo, 2019; p.83). The multidisciplinary team of the UIDT is working on diverse projects; some in approved experimental clinical protocols involving proprietary software and technology, and others involve technology development and innovation in synergy collaboration between technicians and physicians.

The current staff of the UIDT includes three research fellows and one coordinator with employment contracts with the UNAM, but working directly inside the Hospital. Additionally, several medical doctors who work for HGMEL participate in UIDT projects, covering technology development activities for solving specific needs of the Hospital in their clinical areas.

Some of the projects that were developed at the UIDT in its first decade's life were the following: (1) Optimization of the manufacturing process of PMMA craniofacial implants using additive manufacturing techniques, (Ruiz-Huerta et al., 2016). (2) Thermographic utility for diagnosis of micro or macrovascular sickness on diabetic foot, (Fuentes-Oliver et al., 2019). (3) Prostate surgery simulator for training urological surgeons in transurethral resection, (Padilla-Castañeda & Arámbula-Cosıo, 2004;

Arámbula, 2006), (4) Automated system for measuring and diagnosing Glaucoma (IIMAS), (5) design, development and validation of the potential of a UV fluorescence imaging technology for the clinical management of skin wounds (evaluation of diabetic foot), (6) Photothermic response of mammary pathology in ex vivo as probable diagnostic method, (7) Portable corneal topography, (8) Hepatic tissue alterations study of chronic degenerative diseases using opto-thermal techniques implemented using optical fiber, (9) Low cost pressure sensors to measure alterations in footprint in eutrophic children with overweight and obesity, (10) IMU-based wearable systems for human motion assessment in patients with musculoskeletal disorders, such as Ankylosing spondylitis in rheumatism (Martínez-Hernandez et al., 2022) (11) Brachytherapy applicators manufacturing, (12) COVID-19 masks, among others.

Special mention deserves the intensive work that has been carried out with the teaching areas of the Hospital, achieving a very mature relationship in which the Teaching Unit of the Hospital continuously requests projects from the ICAT staff attached to the UIDT.

The ICAT's team has developed several instrumented mannequins and anatomical models for training different anatomical-based clinical techniques, besides several virtual and augmented reality simulators in various surgical specialties, such as orthopedics and neurosurgery, as will be discussed in the next section.

Nowadays, the joint project that is being planned based on the indicators previously discussed in this article is the installation of a Virtual Reality-based Surgery Simulation Laboratory.

7 The Proposal of a Unit of the ICAT's Surgery Simulation Laboratory (SSL)

Over a little more than a decade, the academics, undergraduate, master's, and doctoral students of the ICAT Bioinstrumentation Laboratory have developed technology, software, devices, instruments, mechatronic and robotic systems with medical applications. All these devices and systems constitute the technology basis of the *Surgery Simulation Laboratory* of the ICAT located in the main building of the ICAT in Ciudad Universitaria in Mexico City.

Since 2015 the ICAT's Team has established a close collaboration with the head of the Center for the Improvement of Medical Skills (CPHDM), which is part of the Health Education Directorate of the Hospital, working on projects related to the creation of new models and technologies for the training, acquisition, and improvement of clinical and surgical skills of students, residents, and specialists, improving safety and care for patients.

Derived from this collaboration, since 2016, a virtual reality simulator for neurological vascular microsurgery has been developed in conjunction with the Hospital's Neurosurgery Service, which resulted in the current BACSIM system

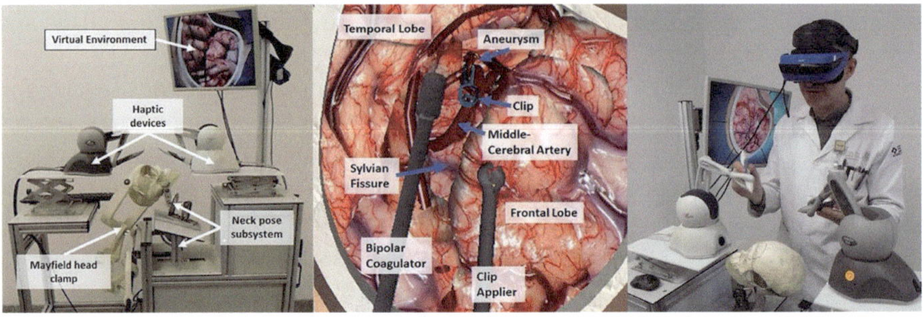

Fig. 7 View of a virtual reality simulation scenario in neurosurgery of the BACSIM system

(Brain Advanced Computer Surgery Simulator) shown in Fig. 7 (Teodoro-Vite et al., 2021), which is a unique system of its kind developed in Mexico and the first installed, in 2019, in a health institution outside a university laboratory, which is in the evaluation and technological maturation phase through the development of new neurosurgical approach scenarios.

From this experience, in 2019, the Bioinstrumentation ICAT's academic group obtained financial support from the Mexico City (CDMX) Secretary of Education, Science, Technology, and Innovation (SECTEI, for its acronym in Spanish), to continue with the development and maturation of the project. Also derived from the BACSIM project, the ICAT's team began to have an important momentum, which led to the development of anatomical teaching models, silicone mannequins, 3D printing models, 3D reconstruction of patient models, virtual reality software and systems, augmented reality, artificial intelligence, and robotics.

From such developments, another representative project developed in collaboration with the Hospital's Algology Department is an anatomical model of the head of a patient electronically instrumented for teaching trigeminal nerve puncture procedures, shown in Fig. 8. This model was used in a practical and theoretical course with residents in 2020–2021, where the theory was taught with the help of a three-dimensional anatomical model by computer, visualized in a group through a virtual reality case and a three-dimensional video projection.

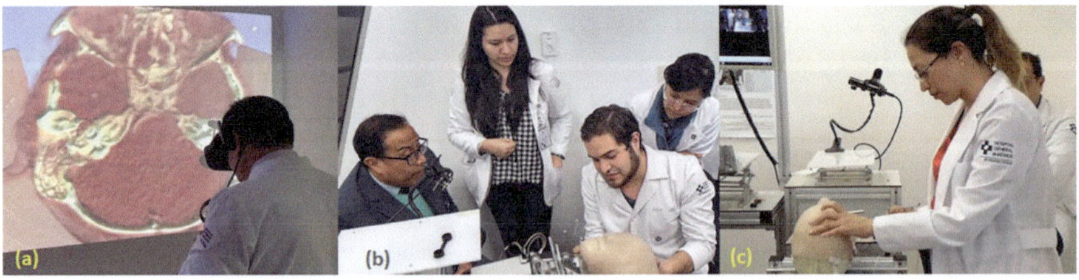

Fig. 8 Teaching strategy of residents of algology at the General Hospital of Mexico "Dr. Eduardo Liceaga." **a** A theoretical session using immersive virtual reality **b** and **c** Practical sessions training in an instrumented anatomical model for blocking the trigeminal nerve

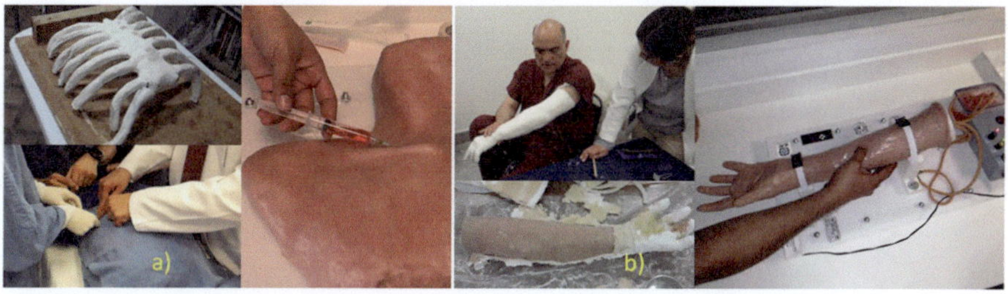

Fig. 9 Instrumented anatomical models **a** A model for central subclavian catheterization **b** a model for peripheral venipuncture procedures

Other projects have been the development of realistic anatomical models and mannequins based on traditional sculpture and molding, besides 3D modeling, image processing, and 3D reconstruction, combined with additive manufacturing techniques commonly known as 3D printing. As a result, various silicone mannequins that faithfully simulate different anatomies, pathologies, and patient models have been produced, mainly requested on demand, specifically for the current needs of the medical service units.

Among some of the most representatives of these projects, it can be mentioned the following (Fig. 9): an electronically instrumented upper arm to train peripheral puncture procedures in Internal Medicine; an electronically instrumented realistic model of a human torso for teaching central subclavian intervention to develop skills of central cannulation, used in 2020 in a course with residents of the Oncology Service.

Besides, the group has developed other virtual and augmented reality systems in clinical areas such as orthopedics and neurosurgery (De Paolis et al., 2022). For example, the SpineNav-AR system which is an augmented reality model for surgical navigation in spine percutaneous procedures (Fig. 10). Some different virtual reality neurosurgical scenarios with haptic feedback for various microsurgical interventions, such as craniotomy and tumor resection (Heredia-Pérez et al., 2018). In addition, an augmented reality model for ventriculostomy training in neurosurgery (Dominguez-Velasco et al., 2023).

With the research and development experiences described and the information obtained from data analysis and technological monitoring of the future of surgical teaching, the idea of establishing a unit of the ICAT's Surgery Simulation Laboratory (SSL) at HGMEL arose.

The initiative to establish a Collaboration Agreement between the ICAT, UNAM, and the Health Education Directorate of the HGM occurred immediately. The scope of the legal instrument was the formalization and the start-up of a laboratory with unique characteristics in Mexico and a pioneer in Latin America to promote the development and research in clinical simulation to improve the clinical training processes of health personnel.

The hypothesis behind this was that the adoption of simulation tools would help students, residents, and medical personnel to develop and acquire skills in various

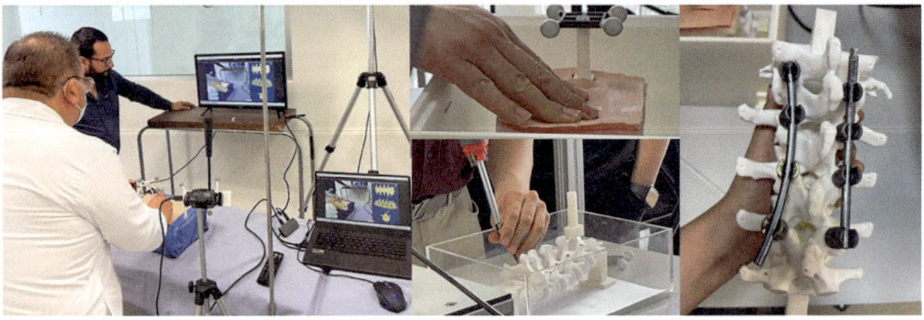

Fig. 10 Views of the SpineNav-AR system, an augmented reality model for training spine percutaneous procedures in orthopedics, such as transpedicular screw placement

complex procedures and interventions in controlled and safe situations, with a variety of clinical cases and levels of demand that generally cannot be practiced intensively in operating rooms with real patients. It will also promote the development of national proprietary technology and the research and evaluation of various diagnostic and treatment techniques in simulated environments in safe conditions without risk to patients.

In summary, the general objective of the Unit of the SSL installed at the HGMEL was to develop research protocols and innovative methods based on engineering technologies and basic science for the study of synthetic and computerized models, improving current clinical and surgical procedures and their teaching methods through the optimal use of simulation tools.

The goal of the Unit is to promote the development of new models and methodologies focused on the training and teaching of medical procedures, which strengthens the training of the personnel of the different services of the Hospital. Furthermore, through the Unit, it will be possible to evaluate and validate new anatomical models and simulators and promote research in new protocols and learning methods based on them. Additionally, the SSL's Unit will encourage development and research into new methods of diagnosis, assistance, and pre and postoperative control that can be extended to various areas of clinical and surgical specialization based on clinical simulation.

The SSL's Unit specific objectives will be:

- Promote the development of research, teaching, and innovation activities in methods and technologies for teaching clinical and surgical abilities and skills among undergraduate and postgraduate students, researchers, and teachers within the Hospital.
- Design, develop, and validate technology based on computerized simulators, mannequins, synthetic tissue models for teaching, quantitative evaluation methods, clinical procedures, surgical interventions, and methods that combine these elements in a curricular way and by learning levels.

- Promote multidisciplinary research, the development and technological transfer of products subject to intellectual property and copyright protection, the development of financed research projects, and the attraction of resources by sponsors.
- Generate innovative technology that results in competitive advantages, such as cost reduction and improvement of teaching processes; promote the development of national technology, as well as reduce risks for patients of the Hospital.
- Help the Hospital modify the morbidity and mortality of the Hospital services through better training of medical staff in safe and controlled environments without risk to the patients.

8 Conclusion

This work shows how the Technology Mining analysis applied for Evidence-Based Foresight helps with the decision making to adjust and improve the current collaboration plan between a public university and the teaching unit of one of the most important federal public hospitals in Mexico. We applied a scientometric and bibliometric information analysis regarding the current digital simulation systems and solid mannequins related to different sicknesses used by medical schools and hospital teaching centers to improve the skills and abilities required by surgeons to diagnose and treat various diseases. The scientometric and bibliometric analysis was used to build a Technology Roadmap driven by authorities of the Applied Science and Technology Institute of the National Autonomous University of Mexico, coordinated with authorities of the Hospital General de Mexico, to decide on creating a Unit of the Surgery Simulation Laboratory (SSL) installed at the Hospital. Potentially, the University's health technologies developed at the Surgery Simulation Lab will be intellectually protected and licensed in the Hospital General de Mexico and other hospitals and R&D institutions of the Mexican Federal Health Sector. Therefore, it was shown that Science, Technology & Innovation (ST&I) data obtained from the scientometric and bibliometric study and the obtained Technology Rout Map are powerful planning tools that help with the future visualization and foresight of the discussed technological field and a solid base for decision making.

Acknowledgements We would like to thank the funding received by the Bioinstrumentation group at ICAT, UNAM for the Surgery Simulation Lab, by Dirección General de Asuntos del Personal Académico, Universidad Nacional Autónoma de México, Mexico, DGAPA PAPIIT TA101422; Secretaria de Educación, Ciencia, Tecnología e Innovación de la Cd. México, for the Project "Laboratorio de investigación y desarrollo de simuladores clínicos y cirugía asistida por computadora" SECTEI 087/2023; and Consejo Nacional de Ciencia y Tecnología, México, CONACYT PCC 319585.

References

Albright, R., & Kappel, T. (2016). Roadmapping in the Corporation. *Research-Technology Management*, pp. 31–40.

Alvarenga dos Santos, T., Orzario Ribeiro, A., & Freitas Silveira, F. (2022). A proposal for a process framework to guide the digital transformation in the R&D Center of a Brazilian Chemical Industry. In *Proceedings of the 7th Brazilian Technology Symposium*, (págs. 502–511).

Arámbula C. F., Padilla C. M. A., P.R. Sevilla M. P. R., (2006), Computer Assisted Prostate Surgery Training. *International Journal of Humanoid Robotics*, *3*(4), pp. 485–498.

Berner, J. E., Ewertz, E., (2018). Theoretical bases of the use of simulation for training in surgery. *Rev. Chilena de Cirugía. 70*(4), pp. 382–388.

Carvalho, M., Fleury, A., & Lopes, A. (2013). An overview of the literature on technology roadmapping (TRM): Contributions and trends. *Technological Forecasting and Social Change*, pp. 1418–1437.

CEPAL. (2021). Tecnologías digitales para un nuevo futuro (LC/TS.2021/43), Comisión Económica para América Latina y el Caribe, Organización de las Naciones Unidas, Santiago, Chile. Available at: chrome-extension://efaidnbmnnnibpcajpcglclefindmkaj/https://repositorio.cepal.org/bitstream/handle/11362/46816/S2000961_es.pdf?sequence=1&isAllowed=y

De Alcantara, D., & Martens, M. (2019). Technology roadmapping (TRM): a systemic review of the literature focusing on models. *Technological Forecasting & Social Change*, pp. 127–138.

De Paolis, L. T., Teodoro-Vite,S., Padilla-Castañeda, M. A., Domínguez-Velasco, C. F., Muscatello, S., Hernández-Valencia, A. F. (2022). An augmented reality platform with hand gestures-based navigation for applications in image-guided surgery: prospective concept evaluation by surgeons. *Int. J. Human–Computer Interaction*, https://doi.org/10.1080/10447318.2021.1926116.

Domínguez-Velasco, C. F., Tello-Mata, I. E., Guinto-Nishimura, G., Martínez-Hernández, A., Alcocer-Barradas, V., Pérez-Lomelí, J. S., Padilla-Castañeda, M. A*. (2023). Augmented reality simulation as training model of ventricular puncture: evidence in the improvement of the quality of punctures. *Int J Med Robot*; e2529. https://doi.org/10.1002/rcs.2529.

Facultad de Medicina, (2022, W/D). Unidad de Investigación en Medicina Experimental. Available at: https://di.facmed.unam.mx/Unidad-de-Investigacion-en-Medicina-Experimental.php.

Fuentes-Oliver, E. I., García-Segundo, C., Solalinde-Vargas, R., Ortiz-Sosa, R., & Serrano-Loyola, R. (2019). Anomalous contra-lateral radiometric asymmetry in the diabetic patient. *Biomedical Physics & Engineering Express, 5*(6), Article 065015. https://doi.org/10.1088/2057-1976/ab49f5

Groenveld, M. (1997). Roadmapping integrates business and technology. research. *Technology Management*, pp. 48–55.

Heredia-Pérez, S., Harada, K., Padilla-Castañeda, M. A., Marinho, M. M., J. Márquez-Flores, J., Mitsuishi, M., (2018). Virtual reality simulation in robotic neurosurgery: evaluating dynamic motion scaling in a master-slave system. *Journal of Medical Robotics and Computer Assisted Surgery*. Aug 16:e1953. https://doi.org/10.1002/rcs.1953.

Hospital General de México., (2022, S/F). Antecedentes Históricos. Disponible en: https://hgm.salud.gob.mx/interna/dirgral/antecedentes.html.

Loyarte, E., Posada, J., Gaines, S., Rajasekharan, S., Olaizola, I., Otaegui, O. J., lorez, J. (2015). Technology roadmapping (TRM) and strategic alignment for applied research center: a case study with methodological contributions. *RD Management*, pp. 474–485.

Manrique-Gutiérrez, G., Motte-García, E., Naveja-Romero, J., Sánchez-Mendiola, M., Gutiérrez-Cirlos, C., (2021). Changes and strategies of medical education in response to the COVID-19 pandemic. *Investigación en Educación Médica. 10*(39), pp. 79–95. https://doi.org/1022201/fm.20075057e.2021.39.21360. Available at: https://www.redalyc.org/journal/3497/349770250010/html/.

Martínez H. A., Padilla C. M. A., Pérez L. J. S, J. Casasola V., Burgos V. R., (2021). Preliminary tests of a inertial measurement units based System for Spine mobility assessment in patients with Ankylosing Spondylitis. *43rd Annual Int. Conf. IEEE Eng. in Medicine & Biology Society (EMBC).* Oct 31-Nov. 4, p. 1–4.

Martínez-Hernández, A., Perez-Lomelí, J. S., Burgos-Vargas, R., & Padilla-Castañeda, M. A. (2022). A wearable system based on multiple magnetic and inertial measurement units for spine mobility assessment: a reliability study for the evaluation of ankylosing spondylitis. *Sensors, 22*, 1332. https://doi.org/10.3390/s22041332

Mastroianni G. N., Busnelli, C. V. N., Huespe. P. E., Dietrich, A., Beskow, A., De Santibañes, M., Pekolj, J., (2020). Changes in the Surgical Training Program in the COVID-19 era. *Rev. Argen Cirug. 112*(2,), pp. 109–118. Available at: https://doi.org/10.25132/raac.v112.n2.1512.es.

Padilla-Castañeda, M. A., & Arámbula-Cosıo, F. (2004). Deformable model of the prostate for TURP surgery simulation. *Computers & Graphics. 28. Nr., 5*, 767–777. https://doi.org/10.1016/j.cag.2004.06.014

Phaal, R., Farrukh, C., & Probert, D. (2001). Technology Roadmapping: linking technology resources to business objectives. Available at Center for Technology Management, University of Cambridge: www.mmd.eng.cam.ac.uk/ctm.

Phaal, R., Farrukh, C., & Probert, D. (2001b). *T-Plan: Fats Start to Technology Roadmapping - planning your route to success.* University of Cambridge.

Probert, D. R., Farrukh, C., & Phaal, R. (2003). Technology roadmapping: linking resources to strategic goals. *Journal of Engineering Manufacture.*

Porras-Hernández, J. D. (2016). Enseñanza y Aprendizaje de la Cirugía. *Inv Ed Med. 5*(20), pp. 261–267. Facultad de Medicina UNAM. Available at: https://doi.org/10.1016/j.riem.2016.05.008.

Ruiz-Huerta, L., Almanza-Arjona, Y. C., Caballero-Ruiz, A., Castro-Espinosa, H. A., Díaz-Aguirre, C. M. and Echevarría y Pérez, E. (2016). CAD and AM-fabricated moulds for fast cranio-maxillofacial implants manufacture. *Rapid Prototyping Journal, 22*(1), pp. 31–39. https://doi.org/10.1108/RPJ-03-2014-0032.

Sánchez-González, P., Oropesa, I., Davis, M., Rodríguez, M., Camba, D., Gómez, E. J. (2018). A new virtual reality based environment for surgical training. XXXVI *Congreso Anual de la Sociedad Española de Ingeniería Biomédica.* Ciudad Real, 21–23 nov, pags. 235- 238. ISBN: 978–84–09–06253–9

Spencer, F. (1978). Teaching and measuring surgical techniques – the technical evaluation of competence. *Bulletin of the American College of Surgeons, 63*(3), 9–12.

Teodoro-Vite, S., Pérez-Lomelí, J.S., Domínguez-Velasco, C. F., Hernández-Valencia, A. F., Capurso-García, M. A., Padilla-Castañeda, M. A. (2021) A high-fidelity hybrid virtual reality simulator of aneurysm clipping repair with brain sylvian fissure exploration for vascular neurosurgery training. *Simulation in Healthcare.* Volume 16 - Issue 4 - p 285–294. https://doi.org/10.1097/SIH.0000000000000489.

Torkington, J., Smith, S. G. T., Reest, B. A., A Darzi, A., (2000). *Ann R Coll Surg Engl*; *82*, pp. 88–94. Available at https://www.ncbi.nlm.nih.gov/pmc/articles/PMC2503534/.

Van Eck, N. J., Waltman, L (2010). VOSviewer: visualizing scientific landscapes. Available from https://www.vosviewer.com.

Vega-González, L. R., García-Segundo, C., (2019). Building productive collaborations: creation of a technological research and development unit (UIDT) of the ICAT in the general hospital of mexico. *Business and Management Studies. 5*(1); ISSN: 2374–5916 E-ISSN: 2374–5924. Redfame Publishing. URL: https://doi.org/10.11114/bms.v5i1.4177.

Using Text Mining Approach to Analyze and Foresight the Energy Research Supply and Demand

Mohammad Rabiei

1 Introduction

The gap between academic research, particularly dissertations and theses, and the needs of society is one of the fundamental challenges of higher education. The rapid growth of universities and the surge in graduate student numbers have elicited criticism from scholars and experts in education and research. This expansion has also led to industry leaders and economic managers expressing dissatisfaction with graduates' limited capabilities in addressing real-world problems and challenges (Mahmoud et al., 2018). Moreover, the prevalence of redundant research is often attributed to the inability of professors and students to identify genuine research problems within various societal sectors. Occasionally, the academic community shifts from "problem-finding" to "problem-creating" (Mahmoud et al., 2018).

Energy is a critical necessity in modern societies, impacting various sectors such as industry, economy, culture, politics, and education. Iran, with its substantial reserves of oil, gas, and fossil fuels, coupled with favorable conditions for developing alternative energy sources like solar power, stands as a key player in the global energy landscape (Zahedi et al., 2022). Conversely, Iran is among the nations with significant energy consumption and heavy expenditures on energy subsidies, accounting for approximately 10% of the world's fossil fuel subsidies. This situation necessitates numerous annual research endeavors in various energy-related fields (Noorollahi et al., 2019). Therefore, there is a pressing need for systematic energy research management. A key strategy in research management is the effective coordination of research supply and demand. This coordination helps to prevent redundant research, reduce the time needed to meet information and research demands, and decrease the waste of research resources (Rabiei et al., 2021).

M. Rabiei (✉)
Information Technology Research Department, Iranian Research Institute for Information Science and Technology (IRANDOC), Tehran, Iran
e-mail: M.rabiei@irandoc.ac.ir

H. da Hora et al. (eds.), *Technology Mining*,
https://doi.org/10.1007/978-3-032-10849-4_7

The 'knowledge gap' refers to the discrepancy between the knowledge necessary for an organization to perform tasks and achieve predetermined goals, and the knowledge acquired from various sources (Case & Given, 2016). This gap can exist in both the current and archived knowledge of an institution, organization, or society, and can encompass both deployable and effective knowledge (McBriar et al., 2003). In other definitions, the knowledge gap pertains to the disparity between existing and encoded knowledge, and the essential knowledge required by a system or individual to attain their objectives (Junnarkar, 1997). In other words, on a macro scale, the term 'Research Gap' denotes the areas that have been overlooked or insufficiently researched within the studies of a community or organization. On a micro scale, it refers to the deficiencies present in the research of a specific subject, which are typically identified in research documents such as articles or theses after a literature review. Therefore, the primary objective of this research is to analyze the gap between the supply of and demand for theses on various topics related to the energy field in Iran, adopting a holistic approach.

The next section will review the related works. Following that, the third section will describe the research methodology. The results will be described, and we will delve into the analysis of the topics concerning the supply and demand sources. This will be followed by a discussion of the gap analysis between these topics. Finally, the research will conclude with some suggestions for policymakers and research managers.

2 Related Works

In recent years, the application of text mining in research management has been observed, encompassing the analysis of research gaps, clustering and classification of research, determination of structures, and text indexing. Numerous studies have utilized text mining tools for research management (Munro & Cairney, 2020; Rabiei et al., 2017, 2021; Tavana et al., 2020). These studies employ database analysis to assess the current state of research and to project future research trends. Furthermore, by examining trends across various research areas, these studies discern and interpret the connections, growth or decline, and the influence of scientific publications within a field on other works in the same domain. They also segment and categorize larger research domains into subfields. Text mining has been shown to reduce the time needed for literature review by 30% to 70%, while missing only 5% of relevant studies (95% recall) (O'Mara-Eves et al., 2015). It enables the examination of a vast corpus of prior studies in a subject area, and search engines aid in extracting new information, relationships, hypotheses, and trends not explicitly stated in the texts. Such extractions would be unfeasible using traditional statistical methods or through manual review of individual resources.

In the field of energy, Munro and Cairney (Munro & Cairney, 2020) conducted a systematic review of the academic literature on energy policy and the policymaking process during sustainable energy transitions. The study concluded that successful sustainable energy transitions necessitate a variety of policy instruments, including regulations, financial incentives, and informational campaigns. The authors also highlighted the importance of effective communication and stakeholder engagement in the policymaking process (Munro & Cairney, 2020). Another systematic review employed a combination of text mining and bibliometric analysis to pinpoint highly prioritized research topics in energy and sustainability studies from 2003 to 2020. This review underscored the significance of fuzzy decision-making techniques in this field. Additionally, it revealed that topics such as energy planning, sustainable energy, renewable energy, and energy policy have become increasingly relevant to researchers in recent years (Tavana et al., 2020).

A recent study explored the latest trends and advancements in wind energy systems, along with the potential role of text mining in reviewing and analyzing wind energy literature. The study determined that text mining techniques are valuable for validating concepts and for serving as a platform to derive requirements for the creation of broadly applicable text mining tools and services in wind energy research. It also pointed out that the primary challenges for wind energy systems include reducing the cost of wind energy, especially offshore, and enhancing the deployment of wind energy while minimizing environmental and social impacts. The article underscored the necessity of a systematic method for reviewing wind energy literature to stay abreast of the rapidly expanding body of knowledge in this area (Eroğlu & Seçkiner, 2016).

Numerous comprehensive studies have been conducted on bibliometrics pertaining to various energy subfields, such as the oil and gas industry (Zhu et al., 2015), electricity (Yang et al., 2021), green energy (Ranganathan & Balasubramani, 2014), solar cells (Dutt, 2016), renewable energy (Romo-Fernández et al., 2013), coal (Yang et al., 2018), and others. The bibliometric analysis of nuclear energy research in Iran from 1990 to 2010 revealed a marked increase in research output, with an average citation rate exceeding five per article. Approximately 93% of the articles were co-authored, with those involving international collaboration garnering more citations than those with only domestic authors. Among Iranian universities, Tehran, Tarbiat Modares, Islamic Azad, and Shiraz have been the most prolific in scientific output. Notwithstanding the critical significance of the energy sector and its research to the nation, bibliometric analyses in Iran's energy field have been scarce (Davarpanah, 2012).

The majority of studies conducted in the field of research gap identification or research prioritization have focused on analyzing the research supply while neglecting the research demand aspect, or they have solely relied on expert opinions to assess the research needs of the entire scientific community. However, expert opinions may differ from the actual needs of the scientific community and researchers. In contrast to similar studies, this research also examines the research demand aspect by analyzing the actual demands expressed by researchers in the search engine of scientific information repository.

3 Methodology

For over half a century, the Iranian Institute for Information Science and Technology (IranDoc) has been engaged in the collection, organization, and dissemination of the country's research resources through the Iranian scientific repository (Ganj[1]), which contains over one million scientific records, more than half of which are theses and dissertations. This study focuses specifically on these two sources and presents both cases under the title of 'thesis.' On average, more than 120,000 searches are conducted daily on this repository. Analyzing this volume of requests can provide valuable information on the supply and demand of research in various fields for researchers, managers, and policymakers in science and technology.

In this research, the collection of theses related to the energy field in Ganj will be considered as the supply dimension. To identify relevant studies in the energy field, subject thesauri, topic modeling, natural language processing (NLP) techniques, and text mining will be applied to extract frequently occurring words. Then, the identified topics in this field will be extracted using a topic modeling approach. The searches conducted by users on Ganj will also be examined and categorized based on the extracted topics. Finally, a matrix of the supply and demand for the identified topics in the energy field will be created, and each topic will be examined from the perspectives of supply and demand.

This study has adopted an integrated method that utilizes text mining techniques from quantitative, bibliographic, and content analysis methods. It also employs natural language processing (NLP) techniques and the use of thesauri to form the bag-of-words and identify texts related to energy. A topic modeling approach is applied to identify the energy topics. The R programming language has been used to develop different parts of this research. Several software packages in R have been utilized, among which the most important are the natural language processing package (Kurt, 2017), the text mining package (Feinerer et al., 2024), and the statistics and machine learning package (Meyer et al., 2023). The temporal scope of the research resources and searches spans from 2018 to 2023.

The research process is illustrated in Fig. 1. In this process, the research steps are executed along two separate yet interrelated paths: the supply (existing theses) and the demand for research (user searches).

In the first step, 10,000 records were randomly selected from the Ganj repository over the past 10 years (1,000 records per year). After preprocessing and utilizing related thesauri, representative words of the energy field were identified. Relevant research sources (25,953 records) and searches related to this field (443,890 records) were also identified, and records of supply and demand were constructed. On the demand side, real user queries were distinguished from search engine robot queries. Following this, topic modeling will be used to determine the topics in this field, including their positions, hot topics, research trends, etc., in both the supply and demand sectors. In the final step, the supply and demand of energy research will be analyzed and visualized.

[1] https://ganj.irandoc.ac.ir.

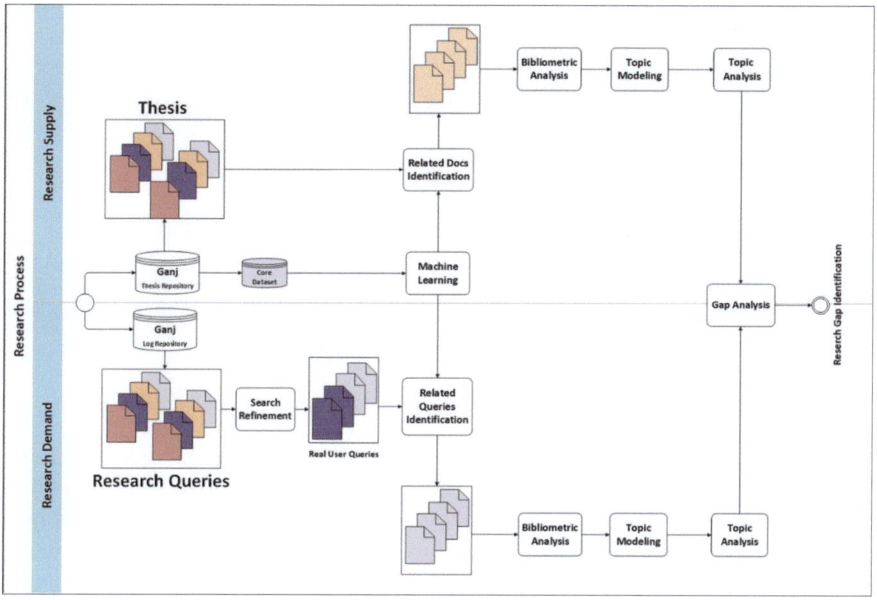

Fig. 1 Research process

4 Topic Modeling

Topic modeling is an effective text analysis method that aims to group a set of different texts into meaningful subgroups, known as topics. This approach accomplishes this task with minimal human intervention, making it attractive and increasingly used in analyzing vast amounts of text. Instead of using predefined titles, this method extracts topic subgroups by specifying only the number of topics, and it also determines the degree of association of each document with each subgroup (Mohr & Bogdanov, 2013). Topic modeling is based on the idea that documents are combinations of topics, where each topic describes a probability distribution over various documents. Conversely, topics are combinations of phrases, and each word describes a probability distribution over topics (Blei et al., 2003).

The LDA algorithm is the most common and reputable form of topic modeling technique, introduced by Blei and his colleagues in 2003 (Blei et al., 2003). The main criticism of this method is its disregard for the order of occurrence of phrases in a text and the order of texts in a corpus. Since each document can cover multiple topics, regardless of the label or topic assigned to it, finalizing the topics of a document is crucial (Chang & Blei, 2009; Lafferty & Blei, 2006). If we denote the importance or weight of a phrase in a document by W and a document contains m phrases, each with its own specific weight and importance for that document, and considering the existence of n documents in a corpus, then LDA can identify the number of k topics in a way that:

$$
\begin{array}{c} \\ \text{Term}_1 \\ \text{Term}_2 \\ \vdots \\ \text{Term}_m \end{array}
\begin{array}{c} \begin{array}{cccc} \text{Doc}_1 & \text{Doc}_2 & \ldots & \text{Doc}_n \end{array} \\
\begin{bmatrix} w_{1,1} & w_{1,2} & \ldots & w_{1,n} \\ w_{2,1} & w_{2,2} & \ldots & w_{2,n} \\ \vdots & \vdots & \ddots & \vdots \\ w_{m,1} & w_{m,2} & \ldots & w_{m,n} \end{bmatrix} \end{array}
\underset{\text{LDA(BoW(Term,Doc), k)}}{\Rightarrow} \tag{1}
$$

$$
\begin{array}{c} \\ \text{Term}_1 \\ \text{Term}_2 \\ \vdots \\ \text{Term}_m \end{array}
\begin{array}{c} \begin{array}{cccc} \text{Topic}_1 & \text{Topic}_2 & \ldots & \text{Topic}_n \end{array} \\
\begin{bmatrix} \beta_{1,1} & \beta_{1,2} & \ldots & \beta_{1,k} \\ \beta_{2,1} & \beta_{2,2} & \ldots & \beta_{2,k} \\ \vdots & \vdots & \ddots & \vdots \\ \beta_{m,1} & \beta_{m,2} & \ldots & \beta_{m,k} \end{bmatrix} \end{array}
*
\begin{array}{c} \\ \text{Topic}_1 \\ \text{Topic}_2 \\ \vdots \\ \text{Topic}_k \end{array}
\begin{array}{c} \begin{array}{cccc} \text{Doc}_1 & \text{Doc}_2 & \ldots & \text{Doc}_n \end{array} \\
\begin{bmatrix} \theta_{1,1} & \theta_{1,2} & \ldots & \theta_{1,n} \\ \theta_{2,1} & \theta_{2,2} & \ldots & \theta_{2,n} \\ \vdots & \vdots & \ddots & \vdots \\ \theta_{k,1} & \theta_{k,2} & \ldots & \theta_{m,k} \end{bmatrix} \end{array}
$$

In the above formula, the first weight matrix (w) represents the m terms (*Term*) in n documents (*Doc*), and this matrix is ultimately equivalent to the matrix product that represents the importance (β) of terms in k topics (*Topic*) and the importance of k topics (θ) in n documents. As a result, topic modeling identifies k topics, where expressions within each topic have their own specific weight. Various methods, such as considering multi-word expressions with the highest weight in a topic or using expert opinion, can be used to select a name or label for that topic. The model's output also specifies the weight of each document within a topic. The primary objective of topic modeling is to identify the topics of documents ($k * n$ matrix). Since the number of topics (k) is much smaller than the number of terms (m) and the number of documents (n), using matrices of size $k * n$ to describe and analyze documents is more comprehensible.

5 Determination of LDA Parameters

The first parameter in LDA is the document-word matrix, which represents the weights of each term in each document. In this study, the number of word occurrences was considered as the weight of each term. Another important input parameter is the value of 'k', which indicates the desired number of topics. There are various methods for determining the number of topics (Blei, 2012; Hall et al., 2008). The advantage of choosing a large number of topics is that it covers all research topic areas, while the advantage of a limited number of topics is that it allows experts and policymakers to have a better understanding and easier interpretation and analysis of the topics (Blei, 2012; Rabiei & Fatahi, 2022; Rabiei et al., 2021). Therefore, it is crucial to select a criterion for determining the number of topics. The perplexity criterion is used for this purpose. Perplexity is a measurement criterion in statistical models used to evaluate the adequacy of a probability distribution or the prediction of a probabilistic model and can be used to compare probabilistic models. In topic modeling, the perplexity of unseen documents is used to evaluate the model, which is a decreasing function of the logarithm of the probability of unseen documents. The lower the value, the better the model's generalizability (Blei et al., 2003). Conversely, a higher value of perplexity signifies that terms in the model have not been appropriately assigned to relevant topics during the learning phase. Given that the implementation of topic modeling on a large number of records (approximately 26,000 records) and the presence of over 12,000 keywords considered as features, the computational cost will be significantly high. This task was accomplished using parallel processing algorithms. The algorithm code in R language is presented in Table 1.

To select an appropriate value for 'k', the model was run using different values of 'k' (from 2 to 50), and the perplexity value was calculated for each run. These values are presented in Fig. 2.

Based on Fig. 2, it is evident that the level of perplexity has significantly decreased at k = 10, and then gradually decreased with a gentle slope. Therefore, the point of intersection of the curve, which is 10, was determined as the optimal value for k. After determining the number of topics (10 topics), topic modeling was performed, and the supplied resources (theses) as well as the search logs were clustered into 10 topics using topic modeling. Subsequently, the R code pertaining to the supplied resources is presented in Table 2. Similarly, the search logs were also modeled into 10 topics.

Table 1 Using parallel processing to determine the perplexity of model for 1:50 topic

```
cl = makeCluster(detectCores()-1) # one core will be free
# pass the variables and functions to cores
clusterExport(cl = cl,c("TermDocMat"))
clusterEvalQ(cl = cl, library(topicmodels))
perplx2 < - as.matrix(parLapply(cl,seq(2,50,1), function(x)
perplexity(LDA(TermDocMat,k = x, method = "VEM"))
stopCluster(cl)
```

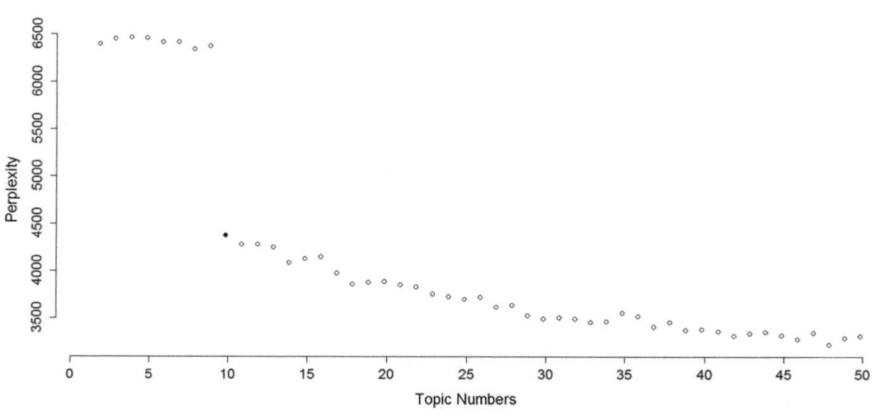

Fig. 2 The perplexity value for different number of topics

Table 2 Topic modeling of supplied thesis to 10 cluster in R

```
EnergyContent < -ThesisEnergy[ThesisEnergy$ID %in% CoreContentEnergy$ID,]
Content < -EnergyContent[,c(1,18)] #1 is ID and 18 is weighted Bag of words
colnames(Content) < -c("doc_id","text")
TCorp < -VCorpus(DataframeSource(Content))
for(i in 1:nrow(Content))
TCorp[[i]]$meta$id < - Content [i,1]
Tdtm < - DocumentTermMatrix(TCorp, control = list(tokenize = CharTok, dictionary =
EnergyTotalTerms, weighting = weightTf))
rowTotals < - apply(Tdtm, 1, sum) #Find the sum of words in each Document
# remove Documents which are not related to any topic
Tdtm < - Tdtm[rowTotals > 0,]
CoreTM50 < -LDA(Tdtm,k = 50, method = "VEM")
```

6 Labeling of Extracted Topics

To better understand the topics, it is recommended to label each topic with a descriptor that represents its content. In this study, the proposed labels for each topic are suggested based on their frequently occurring words. Table 3 displays the extracted topics, their proposed labels, and the frequently occurring phrases for each topic.

7 Results

Since the purpsose of this research is not to analyze the conceptual structure of the energy domain but rather to examine the supply and demand of research topics, the focus will continue on the ten extracted topics.

Frequency of Supplied and Demanded Topics

The frequency of the resources supplied in the energy topics is shown in Fig. 3. An important point in this figure is that since a document may belong to more than one topic, the total resources of the topics are more than the number of resources previously identified for the energy domain.

As shown in Fig. 3, the largest volume of resources has been supplied in the areas of "Technology", "Energy Transmission", "Energy Economy", and "Renewable Energy", while the smallest volume of resources has been allocated to the areas of "Gas", "Oil", and "Solar Energy".

Once the topic modeling of the supplied resources was completed, the resulting model was applied to the data concerning the demanded resources. This time, the demanded resources were categorized into the same ten topics of the energy domain.

Table 3 Extracted topics

Topic	Label	Frequent terms[2]
T 1	Energy economy	Oil revenues, energy management, electricity market, energy consumption reduction, renewable energy, diesel engine, human capital, energy hubs, smart homes, demand response, ...
T 2	Environment	Environmental conservation, lifespan, energy consumption reduction, density functional theory, photovoltaic, renewable energy, alternative energy, energy efficiency, China, plasma, anaerobic digestion, energy optimization, policy, architecture, optimal operation, reverse osmosis, ...
T 3	Technology	Gas hydrates, phase change materials, distribution network, gas turbine, zinc oxide, solar collector, nanoparticles, distributed generation, data panel, microstructure, finite element method, decentralized generation sources, solid oxide fuel cell, surfactant, ...
T 4	Energy transferring	Energy transferring, energy extraction, increasing oil extraction, network lifespan, energy consumption reduction, energy consumption optimization, energy network, super capacitor, project management, piezoelectric, heat exchanger, micro grid, greenhouse gases, energy absorption, mirroring, molecular dynamics simulation, energy loss, clean energy, ...
T 5	Solar energy	Solar energy, energy gap, solar cell, nitrogen, nanostructured materials, energy absorber, Monte Carlo method, wind power plant, gas sensor, wireless sensor networks, efficiency, electrochemistry, ...
T 6	Renewable energy	Renewable energy, energy consumption, solar energy, renewable energy sources, carbon dioxide, Gibbs free energy, energy dissipation, gas turbine, natural gas, biogas, diagenesis, energy management, bioremediation, gas separation, hydrocarbons, renewable energies, scarcity, energy dependency, heat transfer, energy intensity, energy dissipation, ...
T 7	Oil	Crude oil, petroleum industry, gas, subsidies, Russia, petroleum industry, energy consumption, foreign policy, gasoline, diesel, exports, nuclear energy, European Union, graphene, United States, energy resources, legal system, ...
T 8	Energy management	Electricity market, energy consumption, oil, energy loss, energy management, economy, distributed generation sources, job performance, hardship, power grid, energy storage, distributed generation, renewable energy, virtual power plant, cosmology, ...
T 9	Gas	Gas chromatography, corrosion, nanoparticle, gaseous phase, oxidation, nanotechnology, gas synthesis, increased extraction, performance evaluation, gas sensor, increased oil extraction, ...
T 10	Electricity	Electricity, energy consumption, molecular dynamics, solar cell, energy production, thermodynamic properties, renewable energy, air conditioning, electricity supply industry, economic development, ...

Figure 4 illustrates the frequency of searched topics in the energy field within the Ganj repository.

As shown in Fig. 4, the highest demand is in the subjects of "Environment" and "Energy Management", while the lowest demand pertains to the topics of "Energy Economy", "Gas", and "Renewable Energies", respectively.

[2] The original terms are in Persian and their English equivalents are presented in this table.

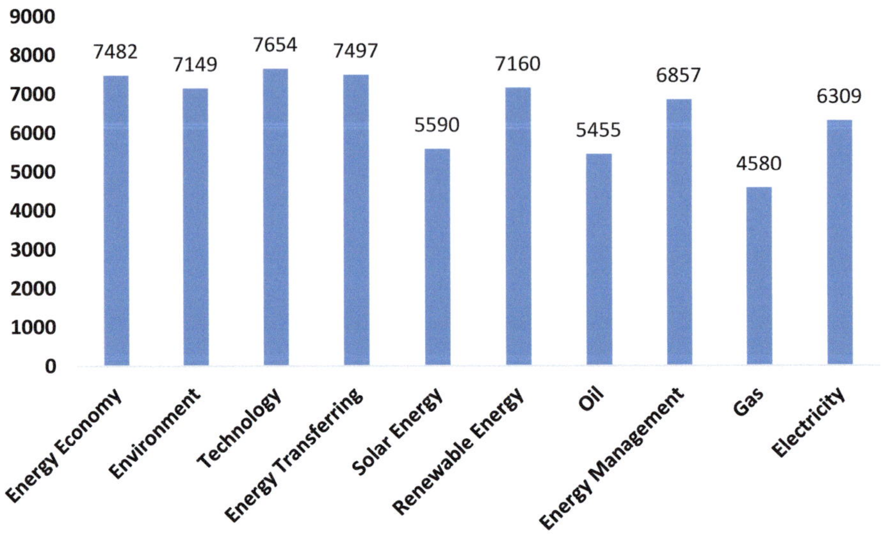

Fig. 3 The frequency of resources supplied in the energy topics

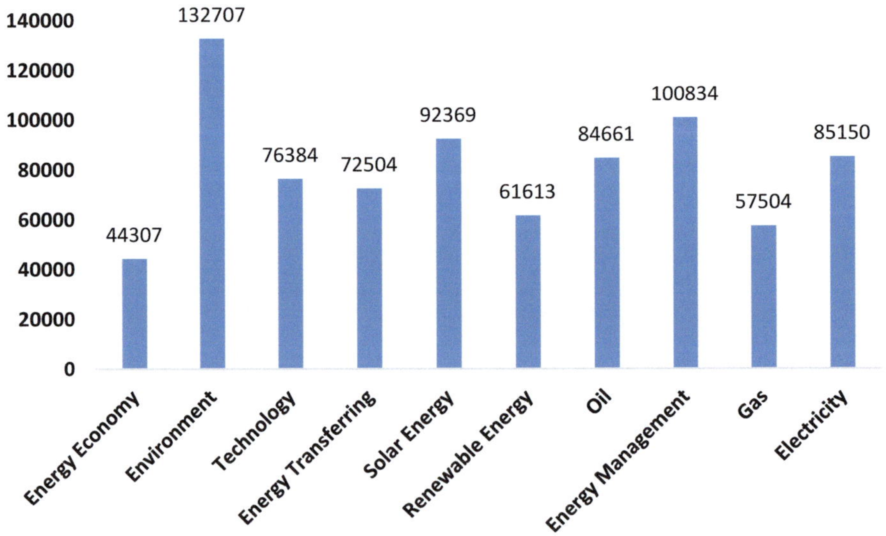

Fig. 4 The frequency of resources demanded in the energy topics

Gp Analysis

To analyze the gap between the research supplied and demanded for the extracted topics, it is necessary to normalize the supply and demand for each topic as a first step. For this purpose, each value of supply and demand was normalized based on their respective minimum and maximum values. The following formula was used for normalization:

$$X'_i = \frac{x_i - Min_i(x)}{Max_i(x) - Min_i(x)} \tag{2}$$

After all values (x_i) were normalized (X'_i) on a scale from zero to one, the topics were mapped based on their positions in the two dimensions of research supply and demand. The horizontal axis representing supply was divided into "high supply"' and "low supply" sections, and the vertical axis representing demand was divided into "high demand" and "low demand" sections. This resulted in a matrix consisting of four quadrants. The value of 0.5 was used as the threshold to distinguish between high and low supply and demand. Figure 5 illustrates the positions of each topic in these four sections.

Based on Fig. 5, the topics of "Environment" and "Energy Management" are considered 'Hot Topics'. Hot topics are assumed to have undergone considerable research and have also received significant attention in user searches. In other words, research on these topics is still being applied, expanded, and extended. A previous study on the supply and demand of the environment observed that the topic of 'Energy' had a high supply (Rabiei et al., 2017), and this study, which focuses on energy, has identified research on energy issues with environmental-related approaches as topics with high supply and demand.

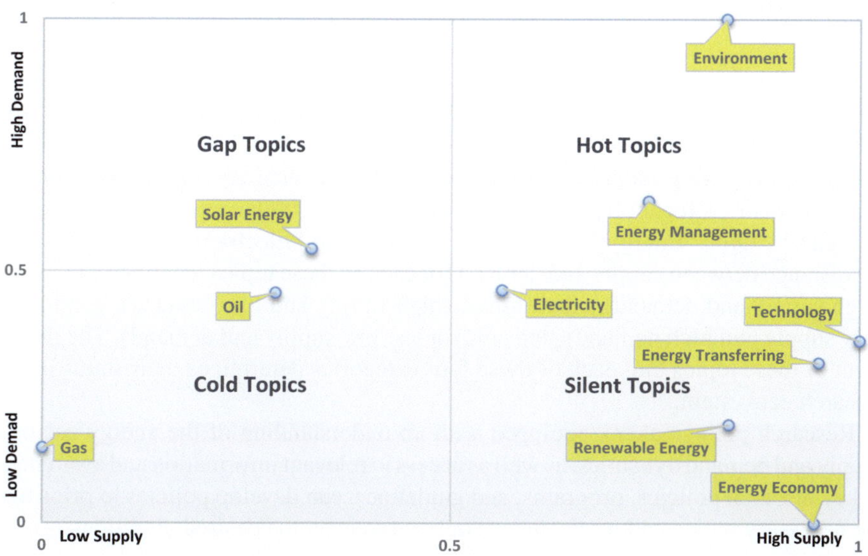

Fig. 5 Supply and demand of research in energy topics

The lower right section of Fig. 5 includes topics chosen as “Silent Topics”, which have been the subject of significant research, but currently, researchers show little interest in using these resources. It appears that some older and saturated research areas are included in this section. The topic of “Electricity” is very close to the previous region and could be classified as a hot topic with some simplification. However, the topics of “Technology”, ’”Energy Economics”, “Energy Transferring’, and “Renewable Energy” are classified as silent topics.

The lower-left section has been precisely identified as the counterpart to hot research topics, labeled “Cold Topics”. “Oil” and “Gas” are located in this section. While “Oil” is somewhat close to the fourth section and has a better status compared to the “Gas” topic, due to the limited number of research studies conducted in this part and the insufficient attention of energy researchers to these topics, it can be predicted that if research policymakers do not motivate researchers for research on the “Gas” topic, it will gradually lose its significance as an independent topic in the future.

The upper-left section encompasses topics that, although they have received attention from researchers and have a high research demand, do not have sufficient research resources available. In other words, despite the high research demand for these topics, users do not have sufficient research resources. Therefore, this part is referred to as “Gap Topics” in the energy domain. “Solar Energy” and to some extent “Oil” are among the topics that face research limitations.

8 Conclusion

In this study, the resources and search queries of users in the Ganj Repository within the field of energy were investigated. The research supply and demand for this field have been monitored. Furthermore, using topic modeling, the topics in the field of energy were identified. These topics were labeled as “Energy Economy”, “Environment”, “Technology”, “Energy Transferring”,”Solar Energy”, “Renewable Energies”, “Oil”, “Energy Management”, “Gas”, and “Electricity”. An analysis of the distance between supply and demand for each of these topics identified hot topics (high supply and demand), silent topics (high supply and low demand), gap topics (low supply and high demand), and cold topics (low supply and demand). The placement of these topics into each of these four categories determines their status in the research ecosystem.

Research policymakers, equipped with an understanding of the energy sector’s supply and demand dynamics, as well as access to relevant information and awareness of key research policies, programs, and guidelines, can develop policies to prioritize the necessary topics and motivate research to focus on the desired directions.

References

Blei, D. M. (2012). Probabilistic topic models. *Communications of the ACM, 55*(4), 77–84.

Blei, D. M., Ng, A. Y., & Jordan, M. I. (2003). Latent dirichlet allocation. *Journal of Machine Learning Research, 3*, 993–1022.

Case, D. O., Case, & Given, L. M. (2016). Looking for information: A survey of research on information seeking, needs, and behavior, Mai, J. -E. ed. 2016: Emerald Group Publishing.

Chang, J., & Blei, D. M. (2009). Relational topic models for document networks. In *International conference on artificial intelligence and statistics*. p. 81–88.

Davarpanah, M. R. (2012) Scientometric analysis of nuclear science and technology research output in Iran. *journal of Scholarly Publishing. 43*(4): p. 421–439.

Dutt, B., & Nikam, K. (2016). Scientometric analysis of global solar cell research.

Eroğlu, Y., & Seçkiner, S. U. (2016). Trend Topic Analysis for Wind Energy Researches: A Data Mining Approach Using Text Mining. *Journal of Technology, 5*(2), 44–58.

Feinerer, I., Hornik, K., & Feinerer, M. I. (2024). Introduction to the tm package text mining in R. CRAN. Availabel at: https://cran.r-project.org/web/packages/tm/vignettes/tm.pdf.

Hall, D., Jurafsky, D., Manning, C. D. (2008). Studying the history of ideas using topic models. In *Proceedings of the conference on empirical methods in natural language processing*. Association for Computational Linguistics. p. 363–371.

Junnarkar, B. (1997). Leveraging collective intellect by building organizational capabilities. *Expert Systems with Applications, 13*(1), 29–40.

Kurt, H. (2017). Natural language processing infrastructure. CRAN. Availabel at: https://cran.r-project.org/web/packages/NLP/NLP.pdf

Lafferty, J. D., & Blei, D. M. (2006). Correlated topic models. In *Advances in neural information processing systems*. p. 147–154.

Mahmoud, Y., et al. (2018). At the nexus of problem-solving and critical research. *Ecology and Society, 23*(4).

McBriar, I., et al. (2003). Risk, gap and strength: Key concepts in knowledge management. *Knowledge-Based Systems, 16*(1), 29–36.

Meyer, D., et al. (2023). Package 'e1071'. CRAN. Available at: https://cran.r-project.org/web/packages/e1071/e1071.pdf.

Mohr, J. W., & Bogdanov, P. (2013). Introduction—Topic models: What they are and why they matter. Poetics. p. 545–569.

Munro, F. R., & Cairney, P. (2020). A systematic review of energy systems: The role of policymaking in sustainable transitions. *Renewable and Sustainable Energy Reviews, 119*, Article 109598.

Noorollahi, Y., et al. (2019). Review of two decade geothermal energy development in Iran, benefits, challenges, and future policy. *Geothermics, 77*, 257–266.

O'Mara-Eves, A., et al. (2015). Using text mining for study identification in systematic reviews: A systematic review of current approaches. *Systematic Reviews, 4*(1), 1–22.

Rabiei, M., et al. (2021). Evolution of IT, management and industrial engineering research: A topic model approach. *Scientia Iranica, 28*(3), 1830–1852.

Rabiei, M., & Fatahi, S. (2022). A new framework for research supply and demand using text analytics. *IEEE Engineering Management Review, 50*(3), 81–95.

Rabiei, M., Hosseini-Motlagh, S.-M., & Haeri, A. (2017). Using text mining techniques for identifying research gaps and priorities: A case study of the environmental science in Iran. *Scientometrics, 110*, 815–842.

Ranganathan, C., & Balasubramani, R. (2014). Scientometric profile of research activities on green energy: An Indian perspective. *International Journal of Research in Applied, Natural and Social Sciences, 2*(1), 23–30.

Romo-Fernández, L. M., Guerrero-Bote, V. P., & Moya-Anegón, F. (2013). Co-word based thematic analysis of renewable energy (1990–2010). *Scientometrics, 97*, 743–765.

Tavana, M., et al. (2020). A review of uncertain decision-making methods in energy management using text mining and data analytics. *Energies, 13*(15), 3947.

Yang, D., et al. (2021). Knowledge mapping in electricity demand forecasting: A scientometric insight. *Frontiers in Energy Research, 9*, Article 771433.
Yang, L., et al. (2018). Mapping of trace elements in coal and ash research based on a bibliometric analysis method spanning 1971–2017. *Minerals, 8*(3), 1–18.
Zahedi, R., Zahedi, A., & Ahmadi, A. (2022). Strategic study for renewable energy policy, optimizations and sustainability in Iran. *Sustainability, 14*(4), 2418.
Zhu, L., et al. (2015). Keywords co-occurrence mapping knowledge domain research base on the theory of Big Data in oil and gas industry. *Scientometrics, 105*, 249–260.

Data-Driven SMEs: An Empirical Analysis of Big Data Analytics Adoption in Germany Using Web Scraping Data

Denilton Luiz Darold and Bernd Ebersberger

1 Introduction

Pervasive computing and high-speed digitalization have led to the generation of massive data pools, enabling the development of new products and services (Geogeghan and Formica, 2018). While traditionally serving a supportive role in decision-making and operational improvements (Wu et al., 2020), data analytics has increasingly assumed a strategic role in fostering innovation (Trabucchi et al., 2018). For this phenomenon, i.e., the combination of data analytics and innovation, scholars in digital innovation have coined the term data-driven innovation (DDI) Rizk et al. (2017). More recently, related concepts like 'data-based innovation' and 'data-driven optimization' have emerged, further expanding the scope of this field Luo (2022). In this study, we encompass all these terms under the broader umbrella of innovation activities focused on investigating underlying technology adoption.

To enable innovation, companies must embrace advanced ICTs, invest in intangible assets, and acquire the necessary skills (OECD, 2021). According to the Oslo Manual, software development and databases are among eight types of activities that firms undertake to innovate, encompassing both product and business process innovations, with Big Data Analytics (BDA) serving as a key enabler in this endeavor (OECD and Eurostat, 2018).

BDA stands out as one of the most critical information and communication technologies (ICTs) employed by firms today, playing a pivotal role in unlocking data's innovative potential (Ajah & Nweke, 2019; Côrte-Real et al., 2019). The importance of BDA lies in its ability to facilitate key organizational processes, such as data

D. L. Darold (✉)
Fraunhofer Institute for Systems and Innovation Research ISI, Karlsruhe, Germany
e-mail: denilton.luiz.darold@isi.fraunhofer.de

D. L. Darold · B. Ebersberger
Department of Innovation Management, University of Hohenheim, Stuttgart, Germany

CIRCLE—Centre for Innovation Research, Lund, Sweden

H. da Hora et al. (eds.), *Technology Mining*,
https://doi.org/10.1007/978-3-032-10849-4_8

cleansing, structuring, and analysis, which are essential for extracting actionable insights (Maroufkhani et al., 2020). Additionally, the rise of digital and connected products has enabled firms to gather extensive data from various sources, predominantly from user interactions (OECD Publishing, 2015). Moreover, the declining data storage and processing costs have made it more feasible for firms to manage large datasets, further promoting the adoption of BDA technologies (OECD, 2020).

Numerous studies have found that BDA influences innovation. Building upon prior research, Niebel et al. (2019) discovered a significant positive relationship between BDA and product innovation in German firms. Qualitative cases show that big data and its analytics and applications can be indicators of organizations' ability to innovate to respond to market opportunities (Wright et al., 2019). Such studies suggest that adopting data-intensive technologies, such as BDA, can facilitate innovation and enhance firms' ability to develop new and improved products.

However, while certain firms have effectively adopted enabling technologies to thrive in the digital era, many have encountered challenges. This challenge is particularly noted among small and medium-sized enterprises (SMEs) (Giovannetti et al., 2020). Empirical studies have consistently demonstrated that larger firms tend to have higher technology adoption rates, often embracing newer and more advanced technologies. In contrast, while making progress in adopting basic ICT technologies, SMEs tend to lag behind when it comes to more complex or integrated technological solutions (OECD, 2021).

SMEs play a critical role in the European economy (Coleman et al., 2016), and this significance is particularly evident in Germany. With SMEs constituting an astonishing 99% of all businesses in the country, they not only serve as major employers, accounting for 55% of the workforce, but also make a substantial contribution of 44% to Germany's GDP (Destatis, 2022). However, despite their economic importance, SMEs have exhibited a relatively slower adoption rate of big data analytics (BDA) compared to their larger counterparts, as observed in prior research (Coleman et al., 2016; Mikalef et al., 2019a). This slowness raises questions about the factors influencing their engagement with BDA and highlights the need for further investigation.

While Germany maintains a significant 10th-place ranking in innovation indicators, its digital competitiveness has faced a decline, falling to the 18th position (WIPO, 2021). In addition, extensive studies have been conducted on the adoption of digital technologies in various countries, including France, Italy, Sweden, and the Netherlands, providing valuable insights into the landscape of technological adoption (OECD, 2021). However, regarding SMEs in Germany, comprehensive empirical evidence at a national level remains limited.

Therefore, there is a research gap that needs to be addressed to understand the determinants of BDA adoption, specifically among SMEs in Germany. Through an empirical analysis, we aim to explore the BDA diffusion and gain a deeper understanding of the determinants of BDA adoption among German knowledge-intensive service firms (KIS) SMEs, answering the following research question:

(RQ1) What Are the Determinants of BDA Adoption in German KIS SMEs?

By using web-scraped text and administrative data, we explore firms' products, services to identify BDA technologies (*what*) and the adopting firms (*who*), followed by an analysis on the determinants of such an adoption. Noteworthy, given the outward-facing nature of our data, we don't investigate internal processes (*how*), i.e., the organizational changes triggered by BDA or how data methods have been employed. Instead, we focus on firms' characteristics in a descriptive analysis of adopters and the potential association with innovation activity by means of patents and trademarks.

In our study, we employ the Technology-Organization-Environment (TOE) framework as the theoretical foundation to analyze the determinants influencing the adoption of Big Data Analytics (BDA). This framework allows us to systematically explore the impact of various factors within organizational and environmental contexts. From an organizational perspective, we will examine how firm age, size, and international ownership may influence BDA adoption. Regarding the environmental context, we will investigate the role of competitive pressure, specifically focusing on market concentration. By analyzing these factors, our study aims to provide a comprehensive understanding of the variables that drive the adoption of BDA. Complementary, we investigate the relationship between BDA technologies and innovative activities by using patent and trademark applications.

In order to operationalize the study, we resort to web scraping data. Commercial databases often lack comprehensive information on firms' technological assets and investments, especially for smaller businesses, hindering large-scale analysis. Following the methodology of previous studies that have successfully employed firm-level web data (Esrock & Leichty, 2000; Oertel & Thommes, 2018; Powell et al., 2016), we interpret a firm's website as expressing its organizational and technological identity. Consequently, we scrape text data from these websites and enrich it by integrating structured data using natural language processing (NLP) techniques. This approach allows us to capture a more nuanced understanding of a firm's technological orientation.

Overall, this study aims to contribute by providing insights extracted from a novel data source, i.e., web data, on a large scale. Our goal is to explore the relationship between organizational characteristics and market conditions, and their influence on the adoption of Big Data Analytics (BDA) and innovation activity among firms. The outcomes may provide input to policy analysis, which for BDA could be seen in two perspectives: 1) boosting firms' adoption of data-related technologies and 2) enabling them to innovate with data by enhancing access to data (OECD, 2021). There are numerous policy initiatives in this field, and Germany launched its Data Strategy, with over 250 policy measures aiming to improve data sharing and its innovative usage (Bundesregierung, 2021). The present study may shed light on the actual impact of those on SMEs.

This paper is structured as follows. Section 2 provides a review of relevant literature and proposed hypotheses. Section 3 describes the methods and data. Section 4 presents the results, followed by a discussion in Sect. 5. Finally, Sect. 6 concludes the paper, and Sect. 6 outlines the agenda for future research.

2 Theoretical Background

The study explores Big Data analytics and is grounded on a TOE framework, commonly used in technology adoption studies.

Big Data Analytics

The rise of digital technologies, digitization, and big data has promoted digital transformation in such a way that many of our social life activities play out in digital spaces. The transformation of subjects, objects, and practices into digital data is called Datafication (Southerton, 2020). Mayer-Schönberger and Cukier (2013) describe it as "the structural translation of everyday activities and interactions into tabulated information that becomes available for analysis process". Datafication is also associated with logic that sees things in the world as data sources and is susceptible to be mined and correlated to obtain insights about human behavior (Southerton, 2020). Furthermore, data is a non-rival good, i.e., it does not dissipate or get used up, so it can be used infinite times in countless different ways while keeping its relevance and, therefore, value. Thus, data can be considered both an infrastructure and capital goods (Geogeghan and Formica, 2018).

Big Data technologies encompass a range of tools and platforms designed to handle the enormous volume, variety, and velocity of data (Qureshi & Gupta, 2014). One of the foundational technologies is distributed computing, which allows for parallel processing of data across multiple nodes in a cluster, enabling efficient and scalable data processing. In addition, NoSQL and columnar databases play a crucial role in handling the diversity of data types and sources in Big Data. They provide flexible and scalable storage solutions for unstructured and semi-structured data, accommodating the diverse nature of data generated from sources such as social media, IoT devices, and sensors (Hu et al., 2014).

Furthermore, advanced analytics and machine learning algorithms enable organizations to extract valuable insights, patterns, and predictions from large and complex datasets (Agarwal & Dhar, 2014). These technologies facilitate exploratory data analysis, statistical modeling, and predictive analytics, empowering organizations to make data-driven decisions, uncover actionable insights, and innovate through data (Hassani et al., 2020; Qureshi & Gupta, 2014).

Technology Adoption

The most popular approaches to technology adoption at the organizational level include the technology-organization-environment (TOE) framework developed by Tornatzky et al. (1990), as well as Roger's (1995) diffusion of innovations (DOI) theory (Oliveira et al., 2011). DOI addresses diffusion by individuals and organizations in a social system, whereas the TOE framework addresses adoption from an organizational perspective, describing three context dimensions that influence technology adoption: technology, organization, and environment. Considering the firm-level perspective of our study, we based it on TOE, exploring the matter according to the dimensions proposed using administrative data.

Previous research has explored adoption of BDA technologies. Agrawal (2017) developed and validated research models to examine the contextual factors influencing BDA adoption. Kang and Kim (2015) and Sun et al. (2018) further explored the factors affecting BDA adoption, with Kang (2015) identifying financial readiness, Information Systems (IS) competence, and industrial pressure as significant factors, and Sun et al. (2018) integrating 26 factors into a TOE framework. Baig et al. (2019) provided a comprehensive review of the theoretical models and influence factors in BDA adoption, highlighting the popularity of the TOE and diffusion of innovations theories.

Regarding SMEs, a range of studies have addressed BDA adoption. Maroufkhani et al. (2020) and Lutfi et al. (2022) further emphasize the role of complexity, uncertainty, and top management support in BDA adoption, with the latter also noting the influence of relative advantage and government support. Mohamed and Weber (2019) underscore the potential of BDA to drive innovation and growth in SMEs, with the latter also discussing the challenges and trends in BDA adoption. Tien et al. (2019) and Tang and Chaveesuk (2021) highlight the importance of technological, organizational, and environmental factors (TOE), with the latter focusing on the manufacturing sector.

TOE

The TOE framework recognizes that the adoption and use of new technologies are influenced by three enterprise contexts: technological, organizational, and environmental (Tornatzky et al., 1990). The TOE model is prevalent in scrutinizing issues related to technology or innovation adoption, such as adopting BDA (Garmaki et al., 2016; Maroufkhani et al., 2020). Although both DOI and TOE are consistent, the TOE framework is considered a more comprehensive instrument for studying technology adoption from an organizational perspective, in the literature (Chen, 2019; Oliveira et al., 2011). Therefore, this study builds on the TOE framework to investigate the adoption determinants of BDA technologies in SME companies in Germany. The focus lies on the firms' organizational characteristics to analyze the determinants of BDA adoption, including variables like size and age, the *rank effects* theorized by (Karshenas & Stoneman, 1993), and also the environment through competitive pressure. The variables influencing the adoption by users, like the ones in technology acceptance models (TAM), are out of the scope, as well as the absorptive capacity, which is presumed given that the scope of analysis is restricted to knowledge-intensive firms.

The coming sub-sections present the TOE dimensions, the proposed variables, and respective grounding and hypotheses. The hypotheses are directional, based on the literature, and submitted to one-tailed tests described in the method's section.

Organization

The organizational characteristics or structures affect the ability to adopt technologies and innovations (Bayo-moriones & Lera-lo, 2007). Two of the most used characteristics are size and age, which are part of the theory of *Rank effects* (Karshenas & Stoneman, 1993), as well as international ownership.

Size: Rogers (1995) argued that the size of the organization directly affects the adoption of innovation. Firm size is widely used in empirical studies on technology adoption as it is easily observed and could serve as a proxy for measuring several phenomena (Geroski, 2000) and has traditionally been found to have a positive effect on the probability of adoption (Gourlay & Pentecost, 2002). Larger firms have fewer financial constraints and are less risk-averse, while smaller firms have more flexibility in adopting new technologies (Haller & Siedschlag, 2014). Several empirical analyses have found a positive correlation between firm size and ICT adoption (Morgan et al., 2006; Thong & Yap, 1996). An OECD study suggests that the higher adoption by larger firms may reflect their needs, such as the diversity of processes that require more advanced ICTs and their capacity to invest in such ICTs and complementary assets (OECD, 2021). In addition, Hollenstein (2004) argued that this relation could be non-linear, with firm size being positively related to early and intensive use of ICT only in firms with up to 200 employees. Considering our scope, focused on SMEs and a relatively new set of data technologies, we hypothesize that:

Hypothesis 1: *The firm size is positively related to BDA adoption.*

Age: Regarding firm age, several empirical studies have identified positive relation to ICT adoption (Dahlke et al., 2022; Haller & Siedschlag, 2014; Hollenstein, 2004). One assumption is that fast-growing firms, new firms, or firms with outdated equipment are more likely to adopt or use new technologies more intensively (Haller & Siedschlag, 2014; Karshenas & Stoneman, 1993). However, when it comes to newer technologies, younger firms are more likely to adopt them (Stel et al., 2005). Thompson et al. (2012) found that younger firms exhibit a higher tendency to adopt newer technologies, as evidenced by the fact that approximately 1 in 3 start-up firms utilize technology that is less than five years old, whereas this proportion drops to around 1 in 10 for firms that have been established for over 13 years. Similarly, when considering a stricter criterion that focuses on technologies or processes available for less than 1 year, a comparable pattern emerges (Thompson et al., 2012).

Therefore, considering the emergence of data-driven technologies, the following hypothesis is established:

***Hypothesis 2**: The firm age is negatively associated with BDA adoption.*

International ownership: Multinational ownership is also commonly evaluated in studies regarding ICT adoption. One argument is that a firm is better informed and able to reduce risk when part of a group (Gourlay & Pentecost, 2002). Another is that an external network influences adoption by raising awareness of innovations (Abrahamson & Rosenkopf, 1997). Westphal et al. (1997) mentioned the pressure to conform to parent firm practices eventually drives adoption. The literature has shown empirical evidence of the positive relationship between international ownership on ICT adoption (Galliano et al., 2001). Therefore, we included it in our analysis, hypothesizing that:

***Hypothesis 3**: International ownership is positively associated with BDA adoption.*

Environment

Environmental context refers to the domain "in which a firm conducts its business—its industry, competitors, access to resources supplied by others, and dealing with the government" (Tornatzky et al., 1990). In this study, we analyze the competitive pressure and its association with BDA adoption, as it is considered relevant in the context of BDA (Lei et al., 2021; Maroufkhani et al., 2022).

Competitive pressure: In the innovation literature, competitive pressure has been recognized as a motivating factor for innovation and the adoption of new technologies (Gatignon & Robertson, 1989). When confronted with intensified competition, firms are more inclined to embrace innovation and incorporate new technologies like ICT to enhance their performance and increase their chances of survival (Haller & Siedschlag, 2014). Numerous studies have demonstrated a positive association between competitive pressure and the adoption of ICT (Hollenstein, 2004; Rao Kowtha and Whai Ip Choon, 2001). However, there's no consensus, as contrasting findings have shown no significant impact of competitive pressure on ICT adoption (Thong & Yap, 1996), or effects may be restricted to specific groups (Bayo-moriones & Lera-lo, 2007). For SMEs, Fosso Wamba et al. (2016) argue that pressure from competitors would promote the adoption of technology among SMEs. Thus, considering the scope of this study, the following hypothesis is proposed:

***Hypothesis 4**: Competitive pressure is positively associated with BDA adoption.*

Technology

The technological dimension is intentionally omitted as the technological indicator—i.e., big data adoption collected via web scraping—is taken as the dependent variable in the analysis of determinants. However, the outcome of our BDA adoption is used as an independent variable to explore the association to innovation activity. Therefore, we review the literature on relations of technology and innovation.

Technology as an Innovation Enabler

The adoption of technology plays a crucial role in enabling data-driven innovations. Information and Communication Technologies (ICTs) are widely recognized as enablers of innovation (Brynjolfsson & Saunders, 2010; Niebel et al., 2019; Santoleri, 2015). Saunders and Tambe (2014), in their analysis of US firms within the IT industry from 1996 to 2012, observed an increasing trend in the utilization of data-related activities (Niebel et al., 2019). Likewise, Brynjolfsson and McElheran (2016) find that the use of data-driven decision-making almost tripled in the US during the period from 2005 to 2010, where the adoption was particularly high in larger firms and in firms with more skilled workers and a higher IT capital stock (Niebel et al., 2019). Galindo-Rueda et al., (2020) also found a positive association between innovation and the adoption of advanced technologies, including ICTs, and business practices in Canadian firms (OECD, 2021). For Germany, Fabritz (2015) provided evidence, using microdata for the positive relationship between ICT investment and firms' propensity to innovate in products.

Specifically on BDA technologies, they have been identified as a key enabler of innovation in various sectors, including business models (Chaudhary et al., 2015; Masenya, 2023), systems and applications (Pan & Yang, 2017), and small and medium-sized enterprises (SMEs) (Masenya, 2023). Luo (2022) states that "Data science, machine learning, artificial intelligence (AI), and computing technologies further empower data-driven innovation". Kopanakis and Mastorakis (2016) argue that techniques and technologies for processing and analyzing "big data" are methods to innovate using data-based decision process. Therefore, we also investigate the relation between BDA adoption and innovation by means of patent applications and trademarks registrations, testing the following hypothesis:

Hypothesis 5: *BDA adoption is positively associated with innovation activity.*

Intellectual Property (IP) has been extensively used as a metric for innovation activity. Patent statistics are widely utilized as indicators of innovative activities, reflecting the inventive performance of countries, regions, and firms (Khan & Dernis, 2006; Pavitt, 2005). Regarding trademarks, Mendonça et al. (2004) and Gotsch and Hipp (2012) both found that trademarks can capture relevant aspects of innovation, particularly in knowledge-intensive services. Ribeiro et al. (2017) argue that the use of trademarks in combination with patents can provide a more comprehensive understanding of innovation. The Oslo Manual 2018 IP data, e.g., patents and trademarks, as one of the main sources for innovation activity. We therefore use these two IP data in our investigation. The overall data and methods are explained next.

3 Data and Method

This study uses four data sources: firms' administrative data from the Orbis database, web text data scraped from the firms' websites, patent applications from European Patent Office (EPO)'s Worldwide Patent Statistical Database (PATSTAT), and trademark fillings from the European Union Intellectual Property Office (EUIPO).

The Orbis database is produced by the Bureau van Dijk (BvD), a leading provider of business intelligence and company information. Orbis covers over 375 million companies worldwide, including both active and inactive companies, and provides information on their financial performance, ownership structure, mergers and acquisitions, and more (Bureau van Dijk, 2022).

The web text data was collected using a Python library *Scrawler* that automatically collects the website data (scraping) and follows inward links mapping the whole domain (crawling). The tool gets the text content from websites and rich metadata, including keywords, language, and depth, measured by the number of sub-levels from the starting URL (SCRAWLER, 2022).

The patent data were collected from the PATSTAT, 2022 spring edition. Trademarks were sourced from EUIPO. The firms collected in both IPR data sources were matched to Orbis in order to create a unified data set and enable the regressions that followed.

The firms' sample collected from the Orbis database selected only SMEs from Germany whose sector (NACE code[1]) is considered knowledge-intensive (KIS) under the Eurostat classification. Eurostat distinguishes between knowledge-intensive services (KIS) and less knowledge-intensive services (LKIS) (EUROSTAT, 2022). While the first group includes NACE sub-sectors 50 to 51, 58 to 63, 64 to 66, 69 to 75, 78, 80, and 84 to 93 (including, for instance, publishing, motion picture, video and television program production, telecommunications, computer-related activities, financial and insurance activities, legal and accounting, head offices, management, architectural and engineering activities, advertising and market research, employment activities, human health, arts, entertainment, recreation, and others), LKIS—further divided into less knowledge-intensive market services and other less knowledge-intensive services—refer to NACE sub-sectors 45 to 47, 49, 52 to 53, 55 to 56, 68, 77, 79, 81, 82, 94 to 96, and 97 to 99 (including activities such as wholesale and retail trade, repair of motor vehicles, land transport, repair of computers and personal goods, etc.). (Schnabl & Zenker, 2013).

The resulting selection criteria applied are the following:

Region: Germany, filtered by the country code in ISO format;
Size: Number of employees less than 250 in the year 2019 (most recent);
Status: Active, according to the weekly status;
NACE code: Ranging from 50 to 93, according to KIS classification. Two codes at the 4-digit level were removed: 7010 and 6420, which refer to "Activities of

[1] NACE Rev. 2—Statistical classification of economic activities https://ec.europa.eu/eurostat/web/nace-rev2.

head offices" and "Activities of holding companies", respectively, i.e., for fiscal or management purposes;
Website: Containing a non-empty string from the website address.

The firms' selection in the Orbis database resulted in 79,038 firms. Next, web scraping was performed for the whole population using the abovementioned scraping tool named *Scrawler*. Given the scale, some thresholds were defined, mainly the limit of 100 pages and the depth of five subdomains/directories per domain. The depth refers to the levels the crawler 'digs' into the domain discovering internal URLs. As the firms' websites are quite diverse, containing unpredictable layouts, data formats, and media files, additional controls are implemented to improve the quality of the data collected and ensure the efficient use of computational resources, avoiding unnecessary scraping.

For the firms' selection, the web scraping successfully scraped 1,706,072 web pages of 64,748 firms. The subsequent data-cleaning procedures can be summarized in the following items:

- Removing the web pages with an HTTP response other than 200 (non-valid)
- Removing the web pages with no valid text information
- Removing the webpages related to data protection URL's data, i.e., whose URL contained the terms: *Impressum, privacy, policy, legal, GDPR*.[2]
- Removing the sentences related to data protection and technical features, like '*cookies*'.

Firms' Selection

The firm identification and classification were made using a dictionary-based method, searching for a determined list of terms (see Appendix C—Keywords list) in the scraped web text data. First, the keywords list was translated into German using the Google Translate API. Next, both the text and keywords were cleaned, removing special characters and reduced to their stemmed form. Finally, the terms were matched to their correspondents in the web text data. The outcome is an overview of technology adoption of data-related technologies, containing the number of occurrences of the technological keywords found on web text both in German and English.

Dictionary Definition

The keywords used in the dictionary method were derived from the reference model for Big Data technologies (Curry et al., 2021) according to five Big Data priority areas: Data Visualization, Data Analytics, Data Processing Architecture, Data Protection, and Data Management. A summarized description is shown in Fig. 1.

[2] General Data Protection Regulation (GDPR).

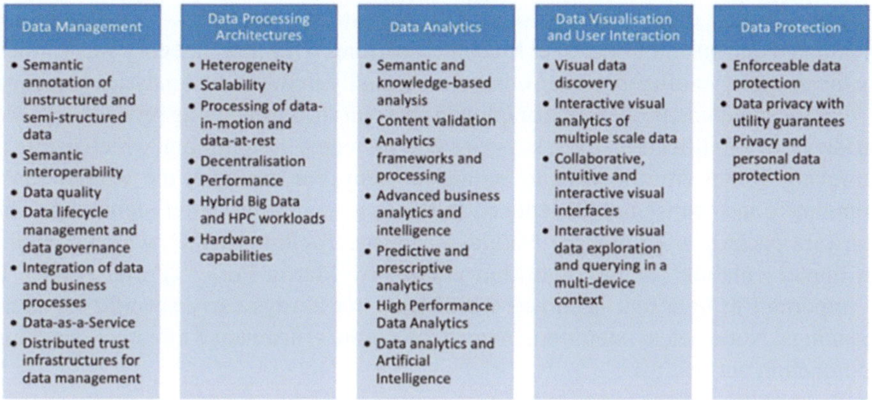

Fig. 1 Big Data priority areas (Curry et al., 2021)

To supplement the keyword list, we've resorted to Search Engine Optimization (SEO) technique. SEO is the set of practices to improve the quality and quantity of website traffic by increasing its visibility and ranking in search engine results pages (SERPs) by targeting specific keywords and phrases that users are searching for and incorporating them into the website's content and metadata. As we collected the data from the firm's websites, it seemed an appropriate way to extend the search and the potential number of matchings.

Based on the SEO rationale, we've used Google Ads Keyword Planner,[3] a web marketing tool designed to discover and research keywords. As for the parameters applied, we kept the default period of one year, i.e., recovering the last year's data and the location, restricted to Germany, therefore matching with the year in which the web scraping was executed and also the geographical scope, as it varies according to the country. The result was manually assessed, and only the relevant words were taken into consideration, removing, for example, vendors' specific terms or slogans, besides other unwanted terms. This particular tool has been used in recent publications under the same goal of keyword generation, although in different thematic fields (Kamí Nski et al., 2019; Yamaguchi et al., 2020).

In addition, the generated list of keywords was supplemented with keywords related to big data skills also collected via web scraping (De Mauro et al., 2018). In this study, whose focus lies on human resources, the required skill set for big data processing is identified and classified. The most popular terms, grouped via Latent Dirichlet Allocation (LDA), provided us with an alternative keyword source to enrich our dictionary. The limitation of this source, as it refers to skills, is that a substantial part is made of actions or behaviors, like *communication and leadership*, or even overly generic, like the *execution.* Nevertheless, when it comes to technologies, it provided a broad and updated set of keywords useful to spot companies and therefore was included in our study.

[3] Google Ads Keyword Planner: https://ads.google.com/home/tools/keyword-planner/

The resulting dictionary contains 175, split into six categories—the five aforementioned and an additional one: Machine Learning. This new category was formed by the artificial intelligence terms that were initially in the Data Analytics category.

For the matching process, we employed an exact string-matching procedure based on keyword identification. Each keyword was subjected to stemming, which involves removing affixes from a word to isolate its stem. For example, the words "run", "running", and "runs" are all reduced to the root word "run" after stemming. The software package used was the Natural Language Toolkit (NLTK), which contains an implementation[4] of the algorithm created by Martin Porter (Porter, 1980). It is important to note that stemmed words may not always carry specific semantic meanings. Nonetheless, stemming allows for a more efficient and effective matching approach in our analysis.

Regression Models Definition

Two regression models were implemented in our study. The first, is a logistic regression analysis to analyze the determinants of BDA adoption. The dataset contains a cross-section of firm-level data, including the BDA adoption indicator generated with the method above and the organizational data collected from the Orbis database. The second, aims to infer the association of BDA adoption and innovation activity by means of patent application and trademarks fillings, from PATSTAT and EUIPO.

Dependent Variables

In the first model, the dependent variable is the BDA-Adoption, a dummy indicating whether a company is an adopter or not. This status is defined based on the occurrence of BDA technologies found on the company's website (BDA_COUNT), with a minimum threshold of three distinct technologies required for a firm to be classified as a BDA adopter. This threshold has proved conservative in manual assessments and represents a relative diversity of data-related technologies, which increases our confidence. Multiple occurrences, even in subdomains, are not considered, contributing, along with the threshold of 100 pages scraped per firm, to mitigate the differences in website size. Of the 175 potential counts, the maximum number of distinct technologies observed on a single firm's website is 29. We opted for a binary variable rather than a continuous index to account for potential discrepancies in how firms present their capabilities—some may inflate their capacities or propositions with buzzwords, while others may neglect technological terms.

In the second model, the dependent variable is also a dummy variable, but for innovation activity by means of patent applications and trademark fillings. The BDA-adoption dummy, the dependent variable in the first model, is then used as

[4] https://www.nltk.org/_modules/nltk/stem/porter.html

the independent one to infer the association between BDA and innovation activity, along with the other independent variables common to both models, explained next.

Independent Variables

As previously mentioned, the independent variables were based on the TOE framework and collected from the Orbis database. The variables reflect the hypotheses previously stated, namely firm size, age, international ownership, and competitive pressure, as follows:

- *Firm size (*NUMBER_OF_EMPLOYEES*)*: The size refers to the logarithm of the number of employees. As hypothesized in **H1**, it is expected that there is a positive relation between firm size and BDA adoption as larger firms face fewer financial constraints and are less risk-averse than smaller firms. However, we focus only on firms with up to 250 employees as larger firms have more diverse business models, which might bias our results. The one-tailed test for this variable presented a p-value of 0.00 and a t-test of 54.176, which suggests a substantial deviation from the null hypothesis and indicates a highly significant relationship between firm size and the dependent variable;
- *Firm age(*FIRM_AGE*)*: The age refers to the logarithm of the number of years since the firm registration. In hypothesis **H2**, we hypothesized that firm age is positively associated with BDA. Although older firms may have more experience in their market and established networks, we argue that younger SMEs would be more willing to adopt new technology and innovation in comparison to older ones. Younger SMEs may have fewer established routines and processes, making it easier for them to experiment with new technologies and ideas. Also, younger SMEs may be more willing to take risks and have fewer sunk costs to protect, which could make them more willing to try out new technologies and innovations. The one-tailed test for this variable resulted in a p-value of 0.00 and a t-test of -12.727, rejecting the null hypothesis and indicating a negative relationship between firm age and the dependent variable;
- *International ownership (*INTL_OWNER*)*: It is a dummy variable set to 1 when the country of the ultimate global owner is different from the domestic owner. The assumption of hypothesis **H3** is that foreign ownership can provide access to technology, knowledge, and managerial skills, which can help firms improve their technological and innovative capabilities. The one-tailed test for this variable showed a p-value of 0.00 and a t-test of 38.899, confirming the positive relationship between international ownership and the dependent variable;
- *Competitive pressure*: Industry concentration via an index with the share of companies from the same sector in the region (IND_CONC). In hypothesis **H4,** we hypothesized that competitive pressure is positively associated with BDA adoption. Companies that face intense competition in their industry may be forced to find new and innovative ways to compete. One way they can do this is by adopting new technologies and innovations that can help them to improve their

products, services, or business processes, in our case BDA. The one-tailed test for this variable conferred a p-value of 0.00 and a t-test of 39.006, which rejects the null hypothesis in favor of the alternative hypothesis that competitive pressure (IND_CONC) positively affects the dependent variable.

Control Variables

As control variables, we have used two commonly used ones: sector and region (Dahlke et al., 2022; Haller & Siedschlag, 2014). Since BDA firms are concentrated in certain sectors and regions (see annex), these variables are used to control for sector- and region-specific effects in the model. The regional control is based on NUTS1-regions in Germany, i.e., the German federal state, and the sector (NACE 1-digit) aggregated by section, e.g., M, J, and K.

4 Results

This section presents the findings, going from the descriptive statistics through the regression analysis and hypothesis validation.

Applying our dictionary-based method, 6343 firms, 9.8% of the whole population, have more than three technological keywords mentioned on their websites, as indicative of BDA adoption. This figure is slightly lower than the estimations found in previous studies. According to OECD, in 2020, on average, in the European Union (27), 34% of large, 21% of medium, and 12% of small firms carried out BDA (OECD, 2021). The distribution of technological keywords can be seen in Fig. 2, and the top 5 keywords in Table 1.

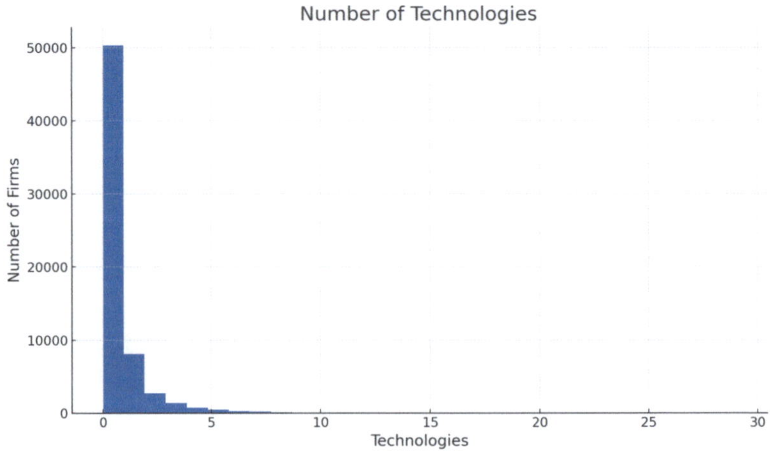

Fig. 2 BDA technologies per firm

Table 1 Top 5 BDA keywords

Keyword	Occurrences*
Artificial Intelligence	3077
Data Analytics	2656
Data Storage	2587
Business intelligence	1657
Data quality	1524

* Absolute occurrences counting

Regarding the categories, data processing takes the lead, followed by data management, data analytics, and machine learning. Data visualization had a small share, and data protection was temporarily ruled out of the study. The data-privacy text on websites is intertwined in such a manner that a dictionary-based identification alone could be misleading. Therefore, we decided to remove it entirely. The resulting dataset, aggregated at the firm level, contains all keywords as dummy variables and the list of technological groups with the sum of distinctive keywords. The overall distribution can be seen in Fig. 3.

Considering the geographical dimension, when grouped by NUTS3, the distribution of BDA firms in Germany is concentrated in big cities, namely Berlin, Munich, Hamburg, Cologne, and Hannover, as depicted in Fig. 4 in the Appendix. Concerning the firms' sector, the most frequent ones are J (Information and communication), M (Professional, scientific, and technical activities), and K (Financial and insurance), according to NACE classification, as shown in Fig. 5 in the Appendix.

For the method validation, a subset of 100 random firms classified as BDA was selected. In a manual check, 92 were flagged as true positives, and only five out of 100 were flagged as false positives. They contain more generic data-related terms but

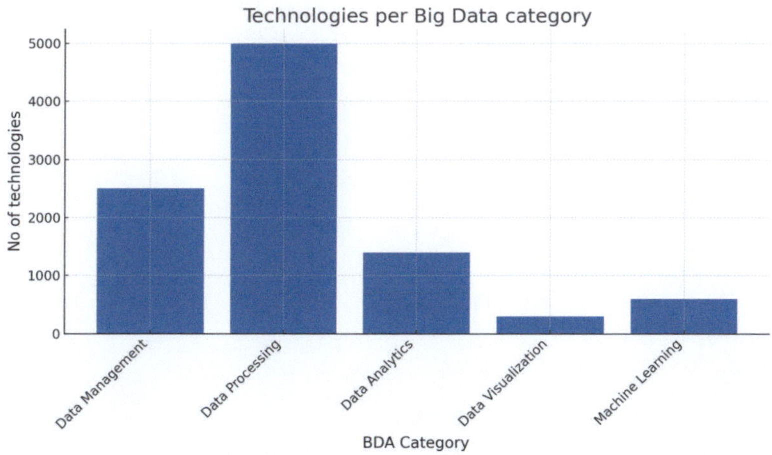

Fig. 3 BDA technologies

clearly don't perform data-related activities. The remaining three were inaccessible, i.e., the website wasn't online anymore.

Descriptive Statistics

The dataset consists of 54,490 observations, providing descriptive statistics for several variables. Regarding *BDA_COUNT*, which represents the count of distinct technologies found on firms' websites, the average count is 0.51 with a standard deviation of 1.51. The range spans from 0 to 29, and the median value is 0. The *FIRM_SIZE* variable, indicating the number of employees in each firm, has an average of 11.69 with a standard deviation of 18.17. The range extends from 1 to 245, and the median value is 5. *FIRM_AGE*, representing the age of the firms in years, has an average of 20.57 with a standard deviation of 12.94. The minimum age is two years, the maximum is 304 years, and the median is 18. Looking at *INTL_OWNER*, approximately 4% of the firms have international ownership, as denoted by the mean value of 0.04. The variable has a standard deviation of 0.19 and a range of 0 to 1. Lastly, the *IND_CONC* index, for market concentration, i.e., the number of firms in the same sector and region, has an average of 12.07 with a standard deviation of 20.85. The range spans from 0 to 100, and the median value is 3.31. The descriptive statistics are shown in Table 2.

The descriptive statistics provided insights into the variables in the dataset, and now we can examine the correlation matrix to explore the relationships between these variables (Table 3).

Table 2 Descriptive statistics

	N_obs	mean	std	max	min	median
BDA_COUNT	54,490	0.51	1.51	29	0	0
FIRM_SIZE	54,490	11.69	18.17	245	1	5
FIRM_AGE	54,490	20.57	12.94	304	2	18
INTL_OWNER	54,490	0.04	0.19	1	0	0
IND_CONC	54,490	12.07	20.85	100	0	3.31

Table 3 Pearson correlation matrix

	BDA_COUNT	FIRM_SIZE	FIRM_AGE	INT_OWN	IND_CONC
BDA_COUNT	1				
FIRM_SIZE	0.146195	1			
FIRM_AGE	−0.025608	0.080596	1		
INTL_OWNER	0.131196	0.083218	−0.024161	1	
IND_CONC	0.10267	−0.009983	−0.072351	0.052037	1

The correlation matrix reveals the interrelationships between the variables in the dataset, shedding light on potential associations among them.

Next, we use the logit model for logistic regression, as previously described. We used the BDA-Adoption indicator (dummy), as the dependent variable and the organizational and context indicators as independent ones. The summary is shown in Table 4.

The analysis supports our hypotheses. First, the analysis confirmed the positive association of FIRM_SIZE to BDA adoption. The coefficient for FIRM_SIZE was found to be 0.305, indicating that an increase in the number of employees is associated with a higher count of technology adoption. Secondly, the hypothesis that firm age is negatively related to BDA adoption was also supported. The coefficient for FIRM_AGE was −0.142, suggesting that older firms tend to have a lower count of technology adoption or, put differently, younger firms are more likely to adopt BDA. Third, the hypothesis regarding international ownership being positively associated with BDA adoption was confirmed. The coefficient for INTL_OWNER was 0.915, indicating that firms with international ownership are more likely to have a higher count of technology adoption. Lastly, the analysis supported the hypothesis that competitive pressure is positively associated with BDA adoption. The coefficient for IND_CONC was 0.010, implying that a higher numerical index combining NACE and NUTS3 classifications is associated with a higher count of technology adoption.

In summary, our hypotheses based on the TOE framework concerning the influence of organizational (size, age, and international ownership) and context (competitive pressure) aspects on BDA adoption by SMEs were supported.

Innovation

Based on the literature supporting the influence of technology adoption on innovation, we created a model to assess the association of BDA adoption and innovation activity, indicated by the existence of patent and trademarks fillings.

Table 4 Logistic regression for BDA adoption

	coef	std err	z
FIRM_SIZE[a]	0.305***	0.009	35.206
FIRM_AGE[b]	−0.142***	0.017	−8.252
INTL_OWNER	0.915***	0.047	19.582
IND_CONC	0.010***	0.000	22.630
Dep. Variable:	BDA-Adoption	**No. Observations:**	54,625
Model:	Logit	**Df Residuals:**	54,619
Method:	MLE	**Pseudo R-squ.:**	0.03962
converged:	TRUE	**Log-Likelihood:**	−28,252

Notes *** $p < 0.01$, ** $p < 0.05$, * $p < 0.1$
Controls are Sector (NACE 2-digit code) and Firm size (number of employees).
[a,b] are log transformed.

For that, we used a dummy variable for innovation activity as dependent variable and BDA Adoption as the independent one, along the organizational and environmental variables. The regression shows a positive association between BDA and innovation, as can be seen in Table 5.

The results of the logistic regression model, as outlined in Table 5, support the hypothesis that BDA adoption, is positively associated with innovation activities. The coefficient for BDA-ADOPTION is significant and positive, indicating a likelihood that firms engaged in BDA are also more likely to engage in innovation activities.

The other independent variables presented similar results to previous regressions, i.e., positive coefficient for FIRM_SIZE, INTL_OWNER and negative for FIRM_AGE, suggesting that the larger, younger, and international firms have a higher propensity for innovation. The difference is the IND_CONC that was found non-significant, indicating that competitive pressure, as measured by industry concentration, might not be a strong determinant of innovation activity in this model.

Overall, the analysis, supported by a relatively high Pseudo R-squared value of 0.1182 compared to related studies (Calvino et al., 2022), underscores the significant influence of both organizational characteristics and technology adoption on innovation activities. This finding aligns with the literature suggesting that technology readiness is critical to fostering an innovative environment within firms.

Robustness Check

We created an alternative model using the number of BDA technologies in a firm instead of the dummy. We employed the negative binomial regression model to examine the relationship between the dependent variable, BDA_COUNT (count of distinct technologies adopted), and the four independent variables. The results, as seen in Table 6, converge to the previous version using a dummy for BDA adoption, i.e., supporting our four hypotheses.

Table 5 Logistic regression for Innovation activity

	coef	std err	z
BDA-ADOPTION	1.526***	0.054	28.478
FIRM_SIZE[a]	0.404***	0.022	18.557
FIRM_AGE[b]	−0.260***	0.041	−6.297
INTL_OWNER	0.332***	0.092	3.602
IND_CONC	0.0003	0.001	0.306
Dep. Variable:	INNOV	**No. Observations:**	54,625
Model:	Logit	**Df Residuals:**	54,613
Method:	MLE	**Pseudo R-squ.:**	0.1182
converged:	TRUE	**Log-Likelihood:**	−6920.7

Notes *** $p < 0.01$, ** $p < 0.05$, * $p < 0.1$
Controls are Sector (NACE 2-digit code) and Firm size (number of employees).
[a,b] are log transformed.

Table 6 Negative binomial regression for BDA adoption

	coef	std err	z
FIRM_SIZE[a]	0.3234***	0.007	46.643
FIRM_AGE[b]	−0.1778***	0.014	−13.135
INTL_OWNER	0.8807***	0.033	27.002
IND_CONC	0.0106***	0.001	20.049
Dep. Variable:	BDA-Adoption	**No. Observations:**	54,625
Model:	GLM	**Df Residuals:**	54,619
Model Family:	NegativeBinomial	**Df Model**	46
Link Function	Log	**Scale**	1
Method	IRS	**Log-Likelihood:**	−45,504
No. Iterations	8	**Deviance**	47,784
converged:	TRUE	**Pseudo R-squ. (CS):**	0.1876

Notes *** $p < 0.01$, ** $p < 0.05$, * $p < 0.1$
Controls are Sector (NACE 2-digit code) and Firm size (number of employees).
[a,b] are log transformed.

Additionally, the same BDA_COUNT was submitted to a Probit Ordered model, where the BDA-adopter was split into categories: non-adopter, basic-adopter, and advanced-adopter. The results are consistent with the Logit and Negative Binomial model, presenting negligible differences for the coefficients.

5 Discussion

The regression models supported the five hypotheses proposed. Based on a sample of German SMEs with up to 250 employees from sectors classified as knowledge-intensive services (KIS), our first hypothesis on the determinants of adoption, firm size, is positively associated with BDA adoption—i.e., the larger the firm, the higher the adoption. This is in line with previous studies mentioned in the background section, which show that the larger the firm, the higher the adoption. The reasons found in the literature are diverse. Larger firms have a higher need for data technologies as they have more complex operations, requiring more information movement and ICT-related tools (Dasgupta et al., 1999). Furthermore, larger firms have more resources to invest, as they have greater access to credit (Bayo-moriones & Lera-lo, 2007) and can reduce the costs involved in the adoption through economies of scale by spreading the fixed cost invested in innovation over a higher number of sales (Cohen & Levin, 1989). Thus, given fewer financial constraints and potentially lower levels of risk aversion, larger firms have the potential to achieve greater profits by adopting new technology, making them better positioned to embrace new technology. In addition, larger firms tend to benefit from innovation complementarities, further enhancing their ability to leverage new technology effectively (Haller & Siedschlag, 2014). Moreover, they have greater human capital, as the relative size of firms is connected to the availability of ICT-related skills (Morgan et al., 2006).

Our second hypothesis was also supported, as firm age is negatively associated with BDA adoption, meaning younger firms tend to adopt more BDA. According to Thompson et al. (2012), newer businesses have a higher probability of utilizing modern technology, primarily because adopting such technology involves lower fixed costs associated with replacing existing technology. As firms mature over time, the utilization of new technology tends to decline. This decline may indicate that these firms no longer require the same level of efficiency as they did during their initial stages, where they focused on overcoming entry barriers or substituting technological advancements for cost reductions based on economies of scale. Another reason, according to the Oslo Manual, is that younger firms may be more agile in implementing change as they are less affected by organizational complexity and inertia (OECD and Eurostat, 2018).

The third hypothesis was also validated regarding the international ownership influence on BDA. Companies owned by a foreign parent company are more likely to adopt BDA. Firms whose headquarters are located abroad might need a higher need for interoperable solutions and also be more exposed to new technologies and market opportunities (Bayo-moriones & Lera-lo, 2007; Dahlke et al., 2022; Hollenstein, 2004).

As for the environment domain, the competitive pressure, measured by industry concentration, based on the share of firms from the same sector (NACE-2-digit) within the same region (NUTS3), positively influences the adoption, therefore supporting the respective hypothesis. The competition has long been associated with technology adoption, as it pressures companies to improve efficiency and explore new business models (Hollenstein, 2004; Oliveira and Martins, 2009). In a highly competitive environment, data-driven technologies may provide a distinct edge over competitors in delivering insights that can help drive both improvements in existing products and services, as well as critical insight toward radically new ones (Mikalef et al., 2020). Moreover, digital technologies may unleash new competitive pressures that call for yet more innovation by tech and non-tech firms alike. New technologies are driving winner-take-all dynamics for an increasing number of industries, pressuring firms to innovate continuously and scale fast so as not to be displaced (Baller et al., 2016).

The fifth hypothesis, on the association of BDA adoption and innovation activity was also supported, in line with previous related studies. Trabucchi and Buganza (2019) defend that big data should be considered not just as a by-product, but as the trigger for the innovation process itself, proposing a data-driven approach to innovation that positions big data as both the enabler and driver of digital innovation processes. According to Niebel et al. (2019), BDA creates novel decision-making possibilities, which are widely believed to support firms' innovation processes. Hao et al. (2019) provide empirical evidence showing that BDA capabilities significantly impact innovation performance, particularly in terms of sales growth and gross margin, underscoring the substantial role of BDA in enhancing sustainable innovation performance. Additionally, Henao-García et al. (2021) discuss how management and process innovation capabilities can foster the development of BDA capabilities, suggesting a close relationship where BDA acts as a facilitator of broader organizational innovation. Finally, Mikalef et al., (2019a, 2019b) explore the role of information governance in enhancing the innovative capabilities driven by BDA, concluding that BDA not only supports incremental and radical innovations but also requires appropriate governance to fully realize its potential.

6 Conclusion

In this paper, we have analyzed the adoption of BDA in Germany, identifying SMEs under the classification of knowledge-intensive service (KIS). The adoption of BDA is measured by big data technologies using web-scraped data.

The technologies considered in this study were based on a reference model developed by Curry et al. (2021), which contains five priority areas: Data Visualization, Data Analytics, Data Processing Architecture, Data Protection, and Data Management. Data Analytics was overrepresented, as it initially included machine learning and artificial intelligence technologies. For that reason, an additional category was created exclusively for Machine Learning.

Our study used multiple data sources: Orbis Database for the firm-level administrative records, where we collected organizational information, and Web data used for the technology scoping. The technological capabilities are rarely available in commercial databases, and the coverage is limited when available. Therefore, data collected from the websites were used to spot technology adoption. The assumption is that firms' websites express their organizational and technological identity. For innovation activity, we collected data from PATSTAT and EUIPO for patents and trademarks, respectively, as a proxy for innovation activity.

The method for identifying the technologies is dictionary-based, searching for predefined BDA terms in the web-scraped data. The search terms were translated into German and underwent a cleaning, normalization, and stemming process. The matching was done via exact string matching, finding correspondences in the text in German, programmatically translated, and also in English. A dummy was created and aggregated into the abovementioned categories for each keyword.

Regarding the determinants for BDA adoption, we based our analysis on the TOE framework, looking into organizational and environmental context indicators. We have developed four hypotheses: (1) that the firm size is positively related to BDA adoption; (2) that the firm age is negatively related to BDA adoption; (3) that international ownership is positively associated with BDA adoption; and (4) that competitive pressure is positively associated with BDA adoption. Complementary, we also hypothesize the BDA adoption has a positive relation to innovation activity, by means of patents and/or trademarks. Our data have supported all of them.

To summarize, our main findings have shown that the propensity of BDA adoption is higher for larger, younger, and foreign-owned firms. In addition, competitive pressure is also positively associated with BDA adoption, i.e., the more competitive the sector a firm is in, the higher the likelihood of adopting BDA technologies.

In conclusion, this study has provided some preliminary insights on the BDA in Germany and the firm characteristics of BDA adopters that help us to understand the determinants for the adoption by knowledge-intensive SMEs. Research on the dynamics of data-driven innovations is increasingly important in mapping the drivers and barriers to developing data-intensive solutions. Policies aimed at promoting innovation and technology adoption in SMEs may need to target younger firms or firms that face intense competition. In addition, policies that encourage foreign investment may also facilitate the adoption of BDA in SMEs. Overall, the findings suggest that the adoption of BDA in SMEs is a complex process that is influenced by a range of internal and external factors, and policymakers should consider such factors to create or evaluate legislative acts to promote innovation and economic growth.

7 Limitations

The interpretation of findings should take into account their limitations. First, it is cross-section analysis and therefore correlational. In order to infer causality, a longitudinal study is required, collecting data from sufficient timespan. Additionally, the BDA indicator is derived from websites and therefore reflects an external, customer-oriented perspective. The innovation model used patents and trademarks, which could bias the result towards firms engaged in hardware-related developments. Lastly, the identification method is dictionary-based, using a matching approach to big data technologies. Such an approach disregards semantic variations; this could be improved by using sentence transformer models, in future work.

8 Future Work

Addressing a limitation of this work, future research could be done by performing a longitudinal data collection, with periodic web scraping, to check how the model hereby presented generalizes over time and how the BDA evolves in a given period. Furthermore, an analysis on the classification of patents and trademarks could shed light on the type of innovations conducted by data-driven SMEs.

Regarding the theoretical frame, other approaches may be applied. One potential theory is the upper echelons theory, which investigates top management's characteristics by analyzing their educational levels, age, online presence, and even personality traits. Additionally, the stakeholder theory or resource dependence theory could also be included in the adoption of BDA.

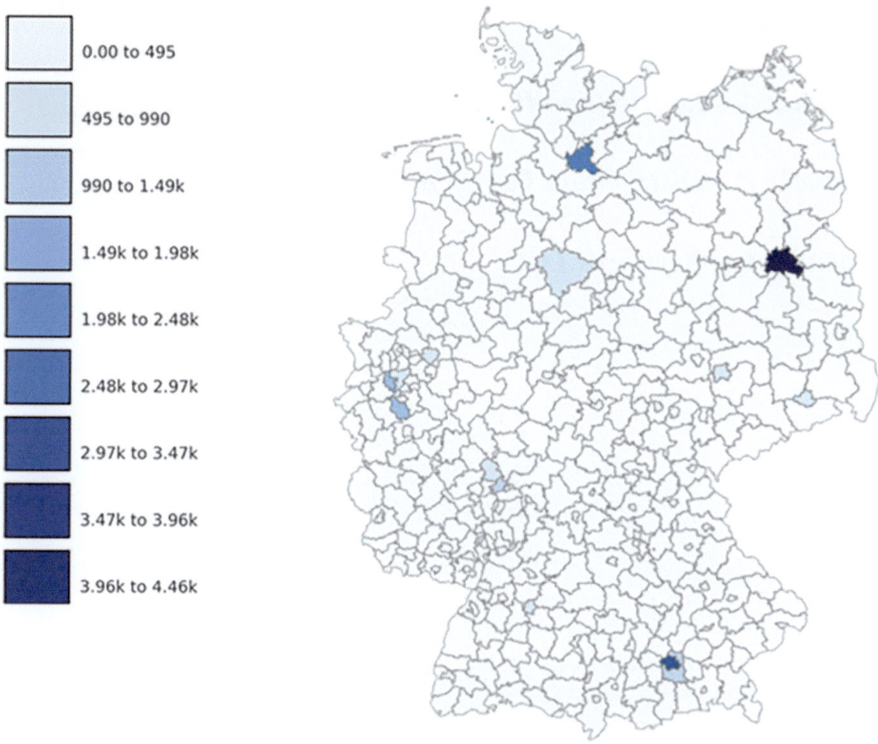

Fig. 4 Geographical distribution of BDA-adopters (NUTS3)

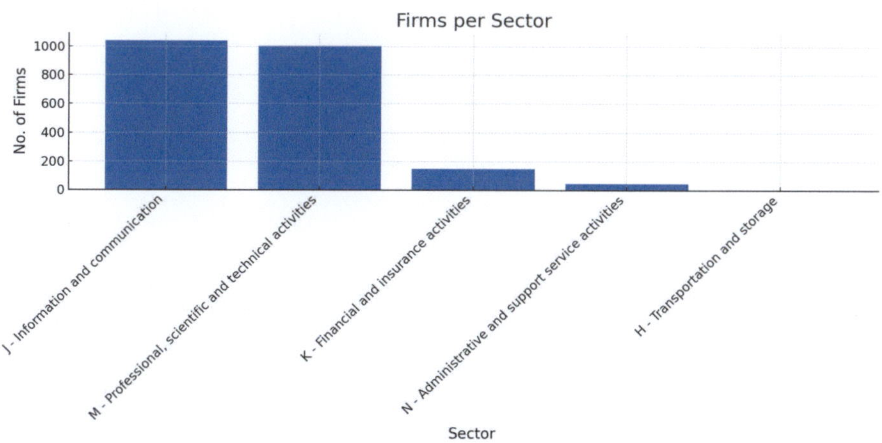

Fig. 5 BDA adopters per sector (NACE section)

Appendix

Appendix A—Distribution Per Region

Appendix B—Distribution Per Sector

Appendix C—Keywords List

See (Table 7).

Table 7 Keywords list

Data management	Data processing	Data analytics	Data protection	Data visualization	Machine learning
Data governance	Extract-transform-load	Data analytics	Data protection	Data visualization	Artificial intelligence
Database management	Data mining	Business analytics	gdpr	Analytics visualization	AI solutions
dmbok	Crips dm	Time series analysis	Data protection officer	Analytics dashboard	Deep learning
Metadata management	Pre-processing	Text analysis	Personal data protection act	Data charting	Neural network
Data lifecycle	Data flow	Big data analytics	Sensitive data	Information visualization	Supervised learning
MDM system	Data ingestion	Advance analytics	Data privacy	DataViz	Unsupervised learning
MDM data	Data extraction	Learning analytics	Data privacy solutions	Data to viz	Ensembled learning
Semantic annotation	Data mining	Customer analytics	Privacy-enhancing technologies	Beautiful data	Teachable machine
Semantic segmentation annotation	ETL data	Marketing analytics	Data anonymization	Visualization tools	TensorFlow
Semantic segmentation label	Data warehouse	Product analytics	Data protection impact assessment	Augmented-reality	Openml
Semantic interoperability	Large scale data	statistical data analysis	k-anonymity	Storytelling with data	Auto ml
Data quality	Data storage	Streaming analytics	Multiparty computation	Graph visualization	ml systems
Data integrity	Microservices scalability	Spatial analytics	Computation	Coordinates plot	Deep ml
Quality analytics	Scalable architecture	Synapse analytics	Risk-based approach	Interactive analytics	Scikit learn
Data lifecycle	Scaling database architecture	Retail analytics	General Data Protection regulation	Multi-device reports	ki machine learning

(continued)

Table 7 (continued)

Data management	Data processing	Data analytics	Data protection	Data visualization	Machine learning
Data lifecycle management	Stream processing	Healthcare analytics		Interactive visual interfaces	Statistical machine learning
Data integration	Batch processing	Data science		Data exploration	Predictive analysis
Integration in business intelligence	Real-time processing	analytics insight		Data querying	Cognitive computing
Business objects etl	Data mesh	data-driven analytics			Interpretable machine
DaaS	Parallel computing	semantic analysis			Learn machine
Data as a service	Distributed system	nlp semantic			Predictive maintenance
Data service	ETL performance	semantic text analysis			Trading bot
Database as a service	Tuning MapReduce jobs	latent semantic analysis			Pattern recognition
Data stream		nlp lsa			Anomaly detection
Data cleansing	Lambda architecture	nltk semantic			
Data analytics	Kappa architecture	Curation of data			
Data factory	Complex Event Processing	Data validation			
Cloud storage	Hadoop	Data normalization			
Distributed cloud infrastructure	Apache Spark	Cross-Industry Standard Process Data Mining			
Distributed computing infrastructure	Apache Flink	CRISP-DM			
Distributed it infrastructure	Apache Kafka	Business intelligence			

(continued)

Table 7 (continued)

Data management	Data processing	Data analytics	Data protection	Data visualization	Machine learning
	Hybrid multi-cloud deployment	Business analytics			
	Internet-of-things	bi software			
	HPCC	Self-service bi			
	High-Performance Computing Cluster	bi analytics			
	Data Analytics supercomputer	Edge analytics			
	Edge computing	Prescriptive analysis			
	Cyber-physical system	High-performance analytics			
	Graphics Processing Unit	hpc data science			
	Object storage infrastructure	hpc data analytics			
	Data storage				
	Data lake				
	Data warehouse				

References

Abrahamson, E., & Rosenkopf, L. (1997). Social network effects on the extent of innovation diffusion: A computer simulation. *Organization Science, 8*, 289–309.

Agarwal, R., & Dhar, V. (2014). Big data, data science, and analytics: The opportunity and challenge for IS research. *Information Systems Research, 25*, 443–448. https://doi.org/10.1287/isre.2014.0546

Agrawal, K. P. (2017). Investigating organizational adoption of big data analytics (BDA) technology. *Americas Conference on Information Systems.*

Ajah, I. A., & Nweke, H. F. (2019). Big data and business analytics: Trends, platforms, success factors and applications. *Big Data and Cognitive Computing, 3*, 1–30. https://doi.org/10.3390/bdcc3020032

Baig, M.I., Shuib, L., & Yadegaridehkordi, E. (2019). Big data adoption: State of the art and research challenges. *Information Processing and Management, 56*. https://doi.org/10.1016/j.ipm.2019.102095.

Baller, S., Dutta, S., & Lanvin, B. (2016). *The global information technology report 2016, lanthanides series determination by various analytical methods.* World Economic Forum. https://doi.org/10.1016/b978-0-12-804704-0.00010-4.

Bayo-moriones, A., & Lera-lo, F. (2007). *A firm-level analysis of determinants of ICT adoption in Spain, 27*, 352–366. https://doi.org/10.1016/j.technovation.2007.01.003

Brynjolfsson, E., & McElheran, K. (2016). The rapid adoption of data-driven decision-making. *The American Economic Review, 106*, 133–139. https://doi.org/10.1257/aer.p20161016

Brynjolfsson, E., & Saunders, A. (2010). *Wired for innovation.* The MIT Press.

Bundesregierung (2021). *Datenstrategie der Bundesregierung.* Available at: https://www.bundesregierung.de/breg-de/suche/datenstrategie-der-bundesregierung-1845632. Accessed 18 Apr 2023.

Bureau van Dijk (2022). *ORBIS database.* Available at: https://www.bvdinfo.com/en-gb/our-products/data/international/orbis. Accessed 7 Apr 2023.

Calvino, F., Samek, L., Squicciarini, M., & Morris, C. (2022). *Identifying and characterising AI adopters: A novel approach based on big data.* https://doi.org/10.1787/154981d7-en

Chaudhary, R., Pandey, J. R., & Pandey, P (2015). Business model innovation through big data. In *2015 International Conference on Green Computing and Internet of Things* pp. 259–263.

Chen, H. (2019). Success factors impacting artificial intelligence adoption: perspective from the telecom industry in China. *ProQuest Dissertations and Theses*, 192. https://doi.org/10.25777/a8q8-gm13.

Cohen, W. M., & Levin, R. C. (1989). Empirical studies of innovation and market structure. In R. Schmalensee, & R. Willig (Eds.), *Handbook of industrial organization.* Elsevier, pp. 1059–1107.

Coleman, S., Göb, R., Manco, G., Pievatolo, A., Tort-Martorell, X., & Reis, M. S. (2016). How can SMEs benefit from big data? *Challenges and a Path Forward*. https://doi.org/10.1002/qre.2008

Côrte-Real, N., Ruivo, P., Oliveira, T., & Popovič, A. (2019). Unlocking the drivers of big data analytics value in firms. *Journal of Business Research, 97*, 160–173. https://doi.org/10.1016/j.jbusres.2018.12.072

Curry, E., Zillner, S., Metzger, A., Berre, A.J., Auer, S., Walshe, R., Despenic, M., Petkovic, M., Roman, D., Waterfeld, W., Seidl, R., Hasan, S., ul Hassan, U., & Ojo, A. (2021). Technical research priorities for big data. *Elements of Big Data Value*, 97–126. https://doi.org/10.1007/978-3-030-68176-0_5.

Dahlke, J., Beck, M., Kinne, J., Lenz, D., Dehghan, R., & Ebersberger, B. (2022). *When is AI adoption contagious? Epidemic effects and relational embeddedness in the inter-firm diffusion of artificial intelligence.*

Dasgupta, S., Ioannidis, A., & Gopalakrishnan, S. (1999). Determinants of information technology adoption. *Journal of Global Information Management.*

De Mauro, A., Greco, M., Grimaldi, M., & Ritala, P. (2018). Human resources for big data professions: A systematic classification of job roles and required skill sets. *Information Processing and Management, 54*, 807–817. https://doi.org/10.1016/j.ipm.2017.05.004

Destatis (2022). Small and medium-sized companies [WWW Document]. URL https://www.destatis.de/DE/Themen/Branchen-Unternehmen/Unternehmen/Kleine-Unternehmen-Mittlere-Unternehmen/_inhalt.html

Esrock, S. L., & Leichty, G. B. (2000). Organization of corporate web pages: Publics and functions. *Public Relations Review*, *26*, 327–344. https://doi.org/10.1016/S0363-8111(00)00051-5.

Eurostat (2022). *Eurostat indicators on high-tech industry and knowledge-intensive services*. Available at: https://ec.europa.eu/eurostat/cache/metadata/Annexes/htec_esms_an3.pdf. Accessed 29 Apr 2023.

Fabritz, N. (2015). ICT as an enabler of innovation. In *Evidence from German microdata ICT as an enabler of innovation evidence from German Microdata*.

Fosso Wamba, S., Gunasekaran, A., Bhattacharya, M., & Dubey, R. (2016). Determinants of RFID adoption intention by SMEs: An empirical investigation. *Production Planning and Control, 27*, 979–990. https://doi.org/10.1080/09537287.2016.1167981

Galliano, D., Roux, P., & Filippi, M. (2001). Organisational and spatial determinants of ICT adoption: The case of French industrial firms. *Environment and Planning A: Economy and Space, 33*, 1643–1663. https://doi.org/10.1068/a3423

Garmaki, M., Boughzala, I., & Wamba, S. F. (2016). The effect of big data analytics capability on firm performance. In *Pacific Asia Conference on Information Systems PACIS 2016—Proceeding*.

Gatignon, H., & Robertson, T. S. (1989). Technology diffusion: An empirical test of competitive effects. *Journal of Marketing, 53*, 35–49. https://doi.org/10.2307/1251523

Geogeghan, M., Formica, P. (2018). *Data-driven innovation*, pp. 123–127. https://doi.org/10.1007/978-3-319-62878-3

Geroski, P. A. (2000). Models of technology diffusion. *Research Policy*, *29*, 603–625. https://doi.org/10.1016/S0048-7333(99)00092-X.

Giovannetti, E., Davies, W., Walsh, L., Willis, R., & Little, J. (2020). *Drivers and barriers for data driven innovation amongst*.

Gotsch, M., & Hipp, C. (2012). Measurement of innovation activities in the knowledge-intensive services industry: A trademark approach. *The Service Industries Journal, 32*, 2167–2184.

Gourlay, A., & Pentecost, E. (2002). The determinants of technology diffusion: Evidence from the UK financial sector. *Manchester School, 70*, 185–203. https://doi.org/10.1111/1467-9957.00291

Haller, S., & Siedschlag, I. (2014). *Determinants of ICT adoption: Evidence from firm-level data*. https://doi.org/10.1080/00036841003724411

Hao, S., Zhang, H., & Song, M. (2019). Big data, big data analytics capability, and sustainable innovation performance. *Sustainability, 11*, 1–15. https://doi.org/10.3390/su11247145

Hassani, H., Beneki, C., Unger, S., Mazinani, M. T., & Yeganegi, M. R. (2020). Text mining in big data analytics. *Big Data and Cognitive Computing, 4*, 1. https://doi.org/10.3390/bdcc4010001

Henao-García, E., Arias-Pérez, J., & Lozada, N. (2021). Fostering big data analytics capability through process innovation: Is management innovation the missing link? *Business Information Review, 38*, 28–39. https://doi.org/10.1177/0266382120984716

Hollenstein, H. (2004). Determinants of the adoption of Information and Communication Technologies (ICT) An empirical analysis based on firm-level data for the Swiss business sector. *Structural Change and Economic Dynamics, 15*, 315–342. https://doi.org/10.1016/j.strueco.2004.01.003

Hu, H., Wen, Y., Chua, T.-S., & Li, X. (2014). Toward scalable systems for big data analytics: A technology tutorial. *IEEE Access, 2*, 652–687. https://doi.org/10.1109/ACCESS.2014.2332453

KamíNski, M., Łoniewski, I., Misera, A., & Marlicz, W. (2019). *Heartburn-related internet searches and trends of interest across six western countries: A four-year retrospective analysis using google ads keyword planner*. https://doi.org/10.3390/ijerph16234591

Kang, D., & Kim, S. (2015). *Process of big data analysis adoption: Defining big data as a new is innovation and examining factors affecting the process*. https://doi.org/10.1109/HICSS.2015.569

Karshenas, M., & Stoneman, P. L. (1993). Rank, stock, order, and epidemic effects in the diffusion of new process technologies: An empirical model. *The Rand Journal of Economics, 24*, 503–528. https://doi.org/10.2307/2555742

Khan, M., & Dernis, H. (2006). Global overview of innovative activities from the patent indicators perspective. *Research Papers in Economics.*

Kopanakis, I., & Mastorakis, G. (2016). Big data in data-driven innovation: The impact in e nterprises ' performance. In *11th International Conference on Management Innovation, Business, and Educational Support Systems*, 257–263.

Lei, Z., Chen, Y., & Lim, M. K. (2021). Modelling and analysis of big data platform group adoption behaviour based on social network analysis. *Technology in Society*, *65*, 101570. https://doi.org/10.1016/j.techsoc.2021.101570.

Luo, J. (2022). *Data-driven innovation: What Is It?*, 1–14.

Lutfi, A., Alsyouf, A., Almaiah, M. A., Alrawad, M., Abdo, A. A. K., Al-Khasawneh, A., Ibrahim, N., & Saad, M. A. (2022). Factors influencing the adoption of big data analytics in the digital transformation era: Case study of Jordanian SMEs. *Sustainability.*

Maroufkhani, P., Wan Ismail, W. K., & Ghobakhloo, M. (2020). Big data analytics adoption model for small and medium enterprises. *Journal of Science and Technology Policy Management, 11*, 171–201. https://doi.org/10.1108/JSTPM-02-2020-0018

Maroufkhani, P., Iranmanesh, M., & Ghobakhloo, M. (2022). Determinants of big data analytics adoption in small and medium-sized enterprises (SMEs). *Industrial Management and Data Systems*. https://doi.org/10.1108/IMDS-11-2021-0695.

Masenya, T. M. (2023). Big data analytics as a game changer for business model innovation in small and medium-sized enterprises in South Africa. *International Journal of Innovation in the Digital Economy, 14*, 1–17. https://doi.org/10.4018/ijide.323136

Mayer-Schönberger, V., & Cukier, K. (2013). *Big data: A revolution that will transform how we live, work and think*. John Murray, London.

Mendonça, S., Pereira, T. S., & Godinho, M. M. (2004). *Trademarks as an indicator of innovation and industrial change.*

Mikalef, P., Boura, M., Lekakos, G., & Krogstie, J. (2019a). Big data analytics capabilities and innovation: The mediating role of dynamic capabilities and moderating effect of the environment. *British Journal of Management, 30*, 272–298. https://doi.org/10.1111/1467-8551.12343

Mikalef, P., Boura, M., Lekakos, G., & Krogstie, J. (2019b). Big data analytics and firm performance: Findings from a mixed-method approach. *Journal of Business Research, 98*, 261–276. https://doi.org/10.1016/j.jbusres.2019.01.044

Mikalef, P., Boura, M., Lekakos, G., & Krogstie, J. (2020). The role of information governance in big data analytics driven innovation. *Information and Management, 57*, Article 103361. https://doi.org/10.1016/j.im.2020.103361

Mohamed, M., & Weber, P. (2019). Trends of digitalization and adoption of big data & analytics among UK SMEs: Analysis and lessons drawn from a case study of 53 SMEs.

Morgan, A., Colebourne, D., & Thomas, B. (2006). The development of ICT advisors for SME businesses: An innovative approach. *Technovation, 26*, 980–987. https://doi.org/10.1016/j.technovation.2005.09.001

Niebel, T., Rasel, F., & Viete, S. (2019). BIG data–BIG gains? Understanding the link between big data analytics and innovation. *Economics of Innovation and New Technology, 28*, 296–316. https://doi.org/10.1080/10438599.2018.1493075

OECD (2020). *OECD digital economy outlook 2020, OECD digital economy outlook 2020*. https://doi.org/10.1787/bb167041-en

OECD (2021). Firms going digital—Tapping into the potential of data for innovation. *OECD Digital Economy*, 1–76.

OECD, Eurostat (2018). Oslo Manual 2018: Guidelines for Collecting, Reporting and Using Data on Innovation, 4th Edition, The Measurement of Scientific, Technological and Innovation Activities, OECD, Handbook of Innovation Indicators and Measurement.

Oertel, S., & Thommes, K. (2018). History as a source of organizational identity creation. *Organization Studies, 39*, 1709–1731. https://doi.org/10.1177/0170840618800112
Oliveira, T., Martins, M. F., & Lisboa, U. N. D. (2011). *Literature review of information technology adoption models at firm level, 14*, 110–121.
Oliveira, T., & Martins, M. F. O. (2009). Determinants of information technology adoption in Portugal. In *ICE-B 2009—International Conference e-Business, Proceeding*, pp. 264–270.
Pan, B., & Yang, Y. (2017). Forecasting destination weekly hotel occupancy with big data. *Journal of Travel Research, 56*, 957–970. https://doi.org/10.1177/0047287516669050
Pavitt, K. (2005). Patent statistics as indicators of innovative activities: Possibilities and problems. *Scientometrics, 7*, 77–99.
Porter, M. F. (1980). An algorithm for suffix stripping. *Program, 14*, 130–137. https://doi.org/10.1108/eb046814
Powell, W. W., Horvath, A., & Brandtner, C. (2016). Click and mortar: Organizations on the web. *Research in Organizational Behavior, 36*, 101–120. https://doi.org/10.1016/j.riob.2016.07.001
Publishing, O. E. C. D. (2015). Data-driven innovation: Big data for growth and well-being. *OECD Publishing, Paris*. https://doi.org/10.1787/9789264229358-en
Qureshi, S. R., & Gupta, A. (2014). Towards efficient big data and data analytics: A review. In *2014 conference on IT in business, industry and government (CSIBIG)*. pp. 1–6. https://doi.org/10.1109/CSIBIG.2014.7056933.
Rao Kowtha, N., & Whai Ip Choon, T. (2001). Determinants of website development: a study of electronic commerce in Singapore. *Information and Management*, *39*, 227–242. https://doi.org/10.1016/S0378-7206(01)00092-1.
Ribeiro, L. C., dos Santos, U. P., & Muzaka, V. (2017). Trademarks as an indicator of innovation: Towards a fuller picture. *Scientometrics, 127*, 481–508.
Rizk, A., Bergvall-Kåreborn, B., & Elragal, A. (2017). Digital service innovation enabled by big data analytics—A review and the way forward. In *Proceedings of the annual Hawaii international conference on system sciences 2017-January*, pp. 1247–1256. https://doi.org/10.24251/hicss.2017.149.
Rogers, M. (1995). Diffusion of innovations. New York 12.
Santoleri, P. (2015). Diversity and intensity of ICT use effects on product innovation: evidence from Chilean micro-data, pp. 1–28.
Saunders, A., & Tambe, P. (2014). Data assets and industry competition: Evidence from 10-K filings. *Ssrn*. https://doi.org/10.2139/ssrn.2537089
Schnabl, E., & Zenker, A. (2013). Statistical classification of knowledge-intensive business services (KIBS) with NACE Rev. 2. evoREG Res. *Note, 25*, 1–11.
SCRAWLER (2022). *GitHub repository*. Available at: https://github.com/dglttr/scrawler/tree/main/scrawler. Accessed 23 Apr 2023.
Southerton, C. (2020). Datafication. In L. A. Schintler, C. L. McNeely (Eds.), *Encyclopedia of Big Data*. Springer International Publishing, Cham, pp. 1–4. https://doi.org/10.1007/978-3-319-32001-4_332-1.
Sun, S., Cegielski, C. G., Jia, L., & Hall, D. J. (2018). Understanding the factors affecting the organizational adoption of big data. *The Journal of Computer Information Systems, 58*, 193–203. https://doi.org/10.1080/08874417.2016.1222891
Tang, W. L., & Chaveesuk, S. (2021). Big data analytics system adoption in SMEs of manufacturing. In: *2021 IEEE 6th International Conference on Computer and Communication Systems*, pp. 28–32.
Thompson, P., Kwong, C. C. Y., & Jones-Evans, D. (2012). *Enterprise education and the adoption of new technologies within small firms, new technology based firms in the new millennium*. Emerald Group Publishing Ltd. https://doi.org/10.1108/s1876-0228(2012)0000009004.
Thong, J. Y. L., & Yap, C. S. (1996). Information technology adoption by small business: An empirical study. *Diffusion and Adoption of Information Technology*, 160–175. https://doi.org/10.1007/978-0-387-34982-4_12.

Tien, E. L., Ali, N. M., Miskon, S., Ahmad, N., & Abdullah, N. S. (2019). Big data analytics adoption model for Malaysian SMEs. *International Conference of Reliable Information and Communication Technology.*

Tornatzky, L. G., Fleischer, M., & Chakrabarti, A. K. (1990). *The processes of technological innovation, issues in organization and management series.* Lexington Books.

Trabucchi, D., & Buganza, T. (2019). Data-driven innovation: Switching the perspective on Big Data. *European Journal of Innovation Management, 22*, 23–40. https://doi.org/10.1108/EJIM-01-2018-0017

Trabucchi, D., Buganza, T., Dell'Era, C., & Pellizzoni, E. (2018). Exploring the inbound and outbound strategies enabled by user generated big data: Evidence from leading smartphone applications. *Creativity and Innovation Management, 27*, 42–55. https://doi.org/10.1111/caim.12241

van Stel, A., Carree, M., & Thurik, R. (2005). The effect of entrepreneurial activity on national economic growth. *Small Business Economics, 24*, 311–321. https://doi.org/10.1007/s11187-005-1996-6

Westphal, J. D., Gulati, R., & Shortell, S. M. (1997). Customization or conformity? An institutional and network perspective on the content and consequences of TQM adoption. *Administrative Science Quarterly, 42*, 366–394. https://doi.org/10.2307/2393924

WIPO (2021). *Global innovation index 2021.* Available at: https://www.wipo.int/global_innovation_index/en/2021/. Accessed 12 Apr 2023.

Wright, L. T., Robin, R., Stone, M., & Aravopoulou, D. E. (2019). Adoption of Big Data Technology for Innovation in B2B Marketing. *Journal of Business-to-Business Marketing, 26*, 281–293. https://doi.org/10.1080/1051712X.2019.1611082

Wu, L., Hitt, L., & Lou, B. (2020). Data analytics, innovation, and firm productivity. *Management Science, 66*, 2017–2039. https://doi.org/10.1287/mnsc.2018.3281

Yamaguchi, Y., Lee, D., Nagai, T., Funamoto, T., Tajima, T., & Chosa, E. (2020). Googling musculoskeletal-related pain and ranking of medical associations' patient information pages: Google ads keyword planner analysis. *Journal of Medical Internet Research, 22*, 1–8. https://doi.org/10.2196/18684

Criminal Investigations, Use of Text Mining and Ontologies

Rogerio Atem de Carvalho and Petterson Faria de Souza

1 Introduction

In the contemporary scenario of criminal investigation, the advent of technology plays a crucial role in collecting and analyzing evidence. In particular, mobile devices, such as smartphones, have become essential sources of information in investigations. Since these store a vast amount of data relevant to the cases, examining texts and the linkage between them offers a new investigative horizon, providing access to messages, call history, location records and a variety of information that can be fundamental in solving crimes. According to Brady et al. (2015), the increasing volume and variety of material from these devices has been well documented and attempts have been made to address the opportunities and problems caused by these two factors.

Criminal investigation is a complex process that involves the collection, analysis and interpretation of evidence to identify and hold criminals accountable. This is a fundamental activity in maintaining public security and in the search for solving crimes. Traditionally, this area of law enforcement has relied on traditional methods of collecting and analyzing evidence. However, with the advent of technology and the increasing amount of information available digitally, the need for more advanced methods of analysis has become evident.

According to Barreto and Wendt (2020), the researcher can be harmed by the lack of breadth in the collection and search of data, generating incomplete knowledge and/or hasty and inaccurate conclusions. Mobile forensics is not just limited to obtaining raw data but involves advanced text mining and natural language processing techniques to extract meaningful insights from these data. These practices allow not only the identification of behavior patterns and connections between individuals,

R. A. de Carvalho (✉) · P. F. de Souza
Instituto Federal de Educação, Ciência e Tecnologia Fluminense, Rio de Janeiro, Brasil
e-mail: r.atem@gsuite.iff.edu.br

P. F. de Souza
e-mail: petterson.faria@gsuite.iff.edu.br

H. da Hora et al. (eds.), *Technology Mining*,
https://doi.org/10.1007/978-3-032-10849-4_9

but also the reconstruction of events, thus strengthening the robustness of criminal investigations.

In this context, the present approach will explore the importance of forensic text examination and data extraction from mobile devices as essential tools for criminal investigators. Furthermore, the crucial role of text mining techniques and semantic web ontologies in the effective interpretation and organization of these data will be discussed, expanding investigative capabilities and accuracy in forensic analysis. By understanding the relevance of these practices, it is possible to advance the understanding and resolution of complex criminal cases, making the most of the information contained on mobile devices to promote justice and public safety.

2 Application of Text Mining in Criminal Investigation

Text analysis emerges as a valuable tool in criminal investigation. This analysis allows investigators to examine messages, documents, and communications to uncover crucial clues and evidence. A central problem in obtaining information from textual data is the identification of fragments of information (Behmer et al., 2019). However, it is important to note that, in many cases, this analysis is carried out manually, consuming significant time and resources, both computational and human resources, since there is a known shortage of Public Security employees. This manual text analysis can be a time-consuming and inefficient process, as investigators must read and interpret large volumes of data, which can lead to errors, omissions, or derail an analysis.

In the age of digital information, forensic text analysis and the application of text mining techniques have proven to be fundamental for criminal investigations. The vast amount of textual data available on electronic devices such as text messages, emails, call logs, and social media offers a rich set of information for investigators. According to Behmer (2019), to extract these fragments from the data, linguistic analysis of the textual data is necessary, which then clarifies the meaning of a given word in the context of the textual data. Text mining emerges as a powerful tool for extracting such data, facilitating the identification of patterns, reconstruction of events, and even the prevention of crimes—such as the following.

Analysis of Messages and Electronic Communications: Text mining allows for the automated analysis of messages and electronic communications retrieved from mobile devices and computers. This analysis involves identifying entities (such as names, dates, locations), classifying content, and identifying relationships between interlocutors. Through this approach, it is possible to reconstruct contact networks, identify possible criminal conspiracies, and even anticipate illicit activities.

Identification of Behavioral and Linguistic Patterns: Text mining is effective in identifying behavioral and linguistic patterns that may be crucial for investigation. Text mining algorithms can detect subtle changes in language patterns, indicating possible changes in behavior that are relevant to the investigation of crimes, such as fraud, threats, or suspicious activity.

Reconstruction of Scenarios and Events: By analyzing texts, such as text messages, emails or social media records, it is possible to reconstruct events, identify chronologies, and establish temporal relationships between different activities. This can help to better understand the context of a crime, determine motivations and validate alibis, significantly contributing to the criminal investigation.

Prevention and Early Intelligence: Text mining also plays an important role in crime prevention. Continuous analysis of textual data can help identify emerging trends, detect potential threats, and provide information for preventative action, allowing authorities to act proactively to prevent crimes from occurring.

Challenges and Ethical Considerations: Although text mining offers numerous advantages in criminal investigation, there are challenges related to privacy, confidentiality, and accuracy of results. Ethics in the collection and use of these data is essential to guarantee the integrity of investigations and respect the individual rights of those involved.

In short, text mining in criminal investigation opens up new perspectives for investigators, allowing for a more in-depth and effective analysis of large volumes of textual data. By integrating this approach with traditional forensic practices, it is possible to improve accuracy in solving cases, contributing to justice and public safety.

3 Use of Ontologies and Semantic Web in Criminal Investigation

The term Semantic Web is defined by Gonçalves and Jacyntho (2020) as an extension of the current Web, providing technologies and standards that allow the addition of explicit formal meaning to published information, which can be understood by machines, making it viable.

The use of ontologies plays a fundamental role in organizing and extracting meaning from textual data, making analysis more effective and accurate. Ontologies are formal representations of knowledge domains. They can be used to structure data and facilitate text analysis. According to Behmer (2019), having textual information represented in an ontology allows us to carry out complex queries and inferences, which can detect patterns of typical criminal actions.

In the context of criminal investigation, ontologies can be used to represent concepts such as crimes, suspects, victims, and evidence. This allows investigators to identify relationships between these concepts and gain deeper insights into cases. The application of ontologies and the semantic web in criminal investigation represents a significant advance in the understanding, organization, and analysis of complex data. These tools offer a structured framework for representing and connecting information, providing a more holistic and contextualized view for researchers.

Data Structuring and Standardization: Ontologies in the semantic web provide a formal structure for representing concepts, their relationships and properties in a given domain. In criminal investigation, this can be applied to standardize terminologies, classify types of crimes, relate individuals, locations, and events, facilitating understanding and analysis of data.

Integration of Data from Different Sources: The semantic web allows the integration of data from different sources and formats. In criminal investigation, this may include data from police records, forensic databases, social media information, among others. By creating a semantic layer over this heterogeneous data, ontologies facilitate correlation and joint analysis, providing a comprehensive view of the investigative landscape.

Semantic Analysis for Pattern Identification: Ontologies allow semantic analysis of data, enabling the identification of patterns, hidden connections and deeper insights. For example, by mapping relationships between people, places and events through an ontology, investigators can identify indirect links between suspicious individuals, places of interest and criminal activities.

Decision-Making Support and Proactive Investigation: By using the semantic web and ontologies, it is possible to create intelligent systems that assist in decision-making during a criminal investigation. Semantic data analysis can generate alerts about suspicious patterns, facilitating the identification of potential threats and contributing to proactive actions by authorities.

Increased Efficiency and Effectiveness in Research: The integration of ontologies and semantic web in investigative practices enables more efficient investigation. Structuring data, identifying patterns, and correlating information helps reduce the time spent manually analyzing large sets of data, making the investigative process more agile and effective.

Challenges and Ethical Considerations: It is important to highlight that the implementation of ontologies and semantic web in criminal investigation faces challenges, such as accuracy in data modeling, privacy and protection of sensitive information, as well as ethical considerations regarding the use and sharing of these data.

In summary, the use of ontologies and semantic web in criminal investigation offers a systematic and structured approach to data analysis, providing investigators with a more comprehensive and accurate understanding of the investigative scenario, thus contributing to the resolution of complex cases and the promotion of justice.

4 Related Work

The **Sleuth Kit**[1] is a collection of command line tools and a C library that allows you to analyze disk images and recover files from them. It is used behind the scenes in Autopsy and many other open source and commercial forensics tools (Carrier, 2023). In **Autopsy**'s documentation,[2] it says that is the leading open-source forensic platform serving the needs of hundreds of thousands of professionals in law enforcement, national security, litigation support, and corporate investigation.

Another well-known framework is the **IBM i2 Analyst's Notebook**. It is a tool that works with Entities, Links, Properties, and Charts. This solution allows the analyst to represent real-world objects and use the power of proprieties, also connect to multi external sources and eliminate a duplicated data.

IPED[3] is the acronym for Digital Evidence Processor and Indexer (from the Portuguese). It is an open source software implemented in Java programming language that can be used to process and analyze digital evidence, often seized at crime scenes by law enforcement or in a corporate investigation by private examiners. This project originally was, and still is, developed by forensic experts from the Brazilian Federal Police since 2012. Although it was always open source, only in 2019 was its code officially published.

The Magnet **AXIOM**[4] is a processor and examiner of digital evidence from mobile, cloud, computer, and vehicle sources allowing the integration of several cases for a unified analysis. It has intuitive analytical tools to automatically surface case-relevant evidence quickly.

MAGNETO, according to Behmer et al. (2019), is the acronym for **M**ultimedia **A**nalysis and correlation en**G**ine for orga**N**ised crime pr**E**vention and inves**T**igati**O**n. An outline of the tool was presented with a specific ontology proposal, use of Speech to Text Transformation and Text Mining.

The proposed tool offers a low cost, simple, yet efficient, alternative to the above tools.

5 Proposed Tool

The biggest challenge of mining texts in cases involving criminal dialogues is the individuals' low level of vocabulary, use of informal words, use of slang and many spelling errors. In some cases, at least one of the individuals does not know how to read or write (illiterate) and audio-only messages are exchanged. In this case, we must transcribe the audio content before applying mining techniques. For example,

[1] https://graphdb.ontotext.com/

[2] https://wordnet.princeton.edu/

[3] https://www.ontotext.com/products/ontotext-refine/

[4] https://sqlitebrowser.org/

Table 1 Example of specific language terms

Term	Portuguese	English
Boat	viatura policial	Police car
Mn	mano	Bro
Vs	visão	Vision
pprt	papo reto	Straight talk
camisa	carga de drogas	Drug shipment

we will use a database from a real case in which the author participated as an Ad hoc Expert.

Dictionary of specific language terms: In dialogue between individuals linked to crime, slang and non-existent vocabulary words are used in order to hinder the automated search for known forensic tools. As a first step, we applied a Python script to handle the messages in order to correct the words known in previous cases, improving the assertiveness rate of the following techniques. Table 1 shows some examples of expressions in specific language.

Boat: Used to warn approaching police vehicles in the area of criminal activity.

mn: Term used between partners or friends.

vs: Term used to ask for attention or priority in dialog.

pprt: Term used to refer to when you are telling the truth, because what has been said may seem dubious. Also used to warn at the beginning of conversations that it will be objective.

camisa: This term is used to try to avoid interception of message content. So they refer to a batch or shipment of drugs as a shirt.

Similarity and spelling correction: the application of this technique helps to correct words that have one or more letters with position displacement, unwanted insertions, and other typing errors or spelling errors.

These examples demonstrate how minor typing errors or spelling variations can significantly alter the meaning of a sentence, potentially leading to misinterpretation or misunderstanding.

Thesaurus verification: to classify subjects it is recommended to use thesauruses to classify messages and content. Princeton University's **Wordnet**[5] project provides a database of semantic annotations.

To map the WhatsApp database, the Semantic Graph Database **GraphDB**1[6] was used and **OntoText Refine**[7] was used to map the data contained in the WhatsApp database. To view the Whatsapp database, **DB Browser for SQLite**3[8] was used. **Linux Mint**[9] was used as the operating system. We use DB Browser to view the

[5] https://linuxmint.com/

[6] http://xmlns.com/foaf/spec/

[7] https://schema.org/

[8] https://www.sleuthkit.org/

[9] https://www.autopsy.com/

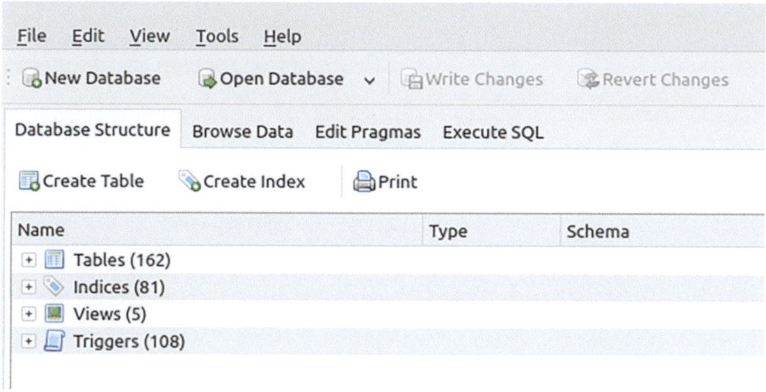

Fig. 1 Overview with DB Browser for SQLite of the msgstore.db database

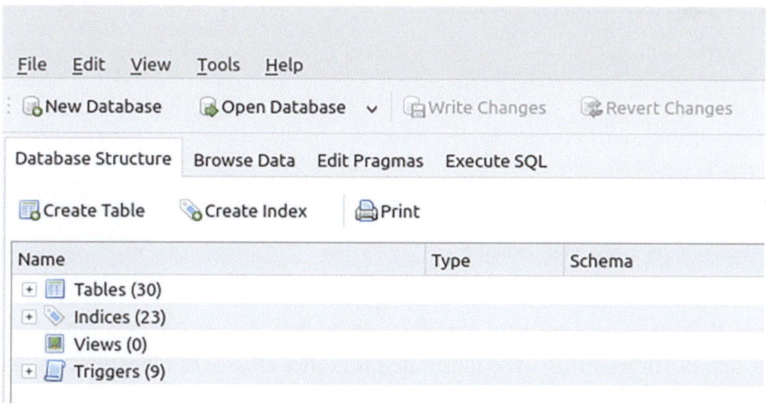

Fig. 2 Overview with DB Browser for SQLite of the wa.db database

tables of the WhatsApp databases of interest (wa.db and msgstore.db). Normally these bases are not accessible by users without privileges on the system, whether Android or iOS. To obtain the files, we use the technique known as "apk downgrade", where a version of WhatsApp with a known vulnerability is installed and thus elevates the privilege to copy the database and its encryption key (Figs. 1 and 2).

Next, we proceed with the creation of a database in GraphDB. And connect Refine to the two WhatsApp databases of interest (wa.db and msgstore.db). We initially applied a mapping of the fields using the Foaf1[10] and schema.org2[11] Ontologies as shown in Figs. 3 and 4.

The message table was also mapped for future text processing through text mining.

[10] https://www.github.com/sepinf-inc

[11] https://www.magnetforensics.com/products/magnet-axiom

Fig. 3 RDF Schema msgstore_chat

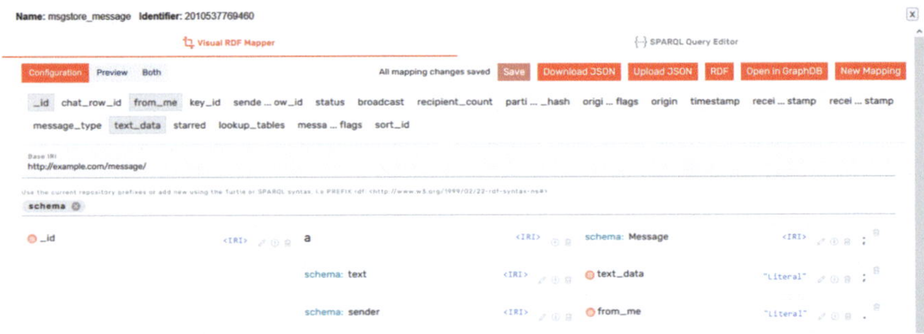

Fig. 4 msgstore_message RDF Schema

The tool must be operable by members of the law forces who are not necessarily experts, that is, for usability to be facilitated for other classes of investigators. Another major challenge of working with messages is that it is a common practice for users to delete conversations, which makes retrieving and analyzing this content difficult. And the use of audio to avoid textual searches requires us to perform an automated transcription of the audio files and then apply text mining techniques. To overcome this problem of deleted messages, we use a merge operation with previous backup databases and the current database. In many cases, we can recover crucial information to close cases. However, with each modification or update of the database via Whatsapp, the scripts generally need to be updated (Fig. 5).

Figure 6 presents the results of a SPARQL query, highlighting the effectiveness of using ontologies in text mining for criminal investigations. The query returns a list of individuals (foaf:Person) extracted from a WhatsApp conversation database that was previously mapped and structured. The application of text mining and ontologies in criminal investigations is crucial for transforming unstructured data into valuable and organized information. As highlighted by Manning et al., (2008), text mining enables the automatic extraction of meaningful information from large volumes of textual data. In a criminal investigation context, as discussed by Carter (2009), these techniques allow for the identification of behavioral patterns and the prediction of

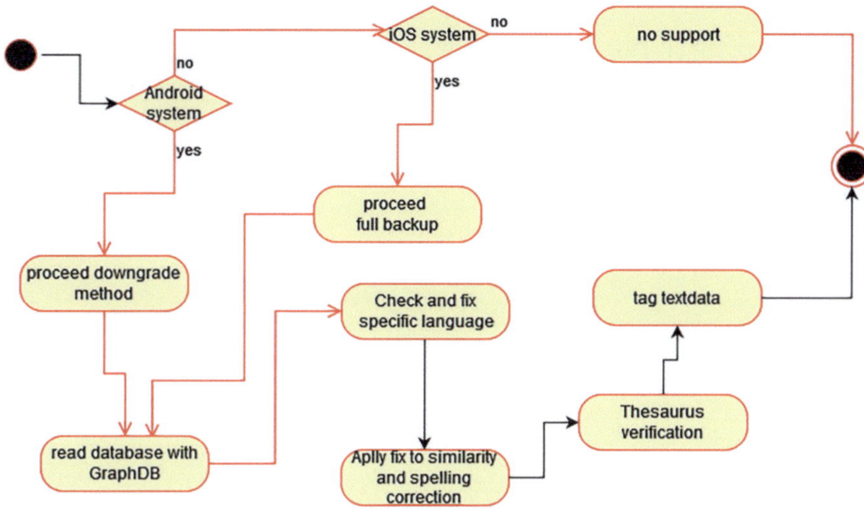

Fig. 5 Flow from extraction to text exam

criminal activities. Furthermore, the use of ontologies, as explained by Hendler et al. (2009), facilitates the modeling of complex communications into a semantic structure that can be easily analyzed and inferred. In this study, we exemplify these approaches through a SPARQL query on a WhatsApp conversation database, demonstrating how these tools can be practically applied to identify crucial connections in contact networks. By aligning our work with the tech mining community, as suggested by Porter and Cunningham (2005), we strengthen the relevance and applicability of our approach within the broader context of technology mining and data analysis.

Furthermore, recent studies in the field of computational linguistics and natural language processing have emphasized the significance of text mining techniques in various domains, particularly in forensic data analysis. For instance, Behmer et al. (2019) proposed an Ontology Population Framework for instantiating heterogeneous forensic data modalities, highlighting the importance of structured data representation in criminal investigations. Similarly, Gonçalves and Jacyntho (2020) introduced a method for semantic Linked Data publication of conventional databases, offering insights into the integration of semantic technologies in forensic data management. These contributions from the academic community, published in reputable journals such as *Artificial Intelligence Applications* and *Innovations and Transformation*, underscore the growing interest and advancements in leveraging text mining and ontologies for criminal investigations.

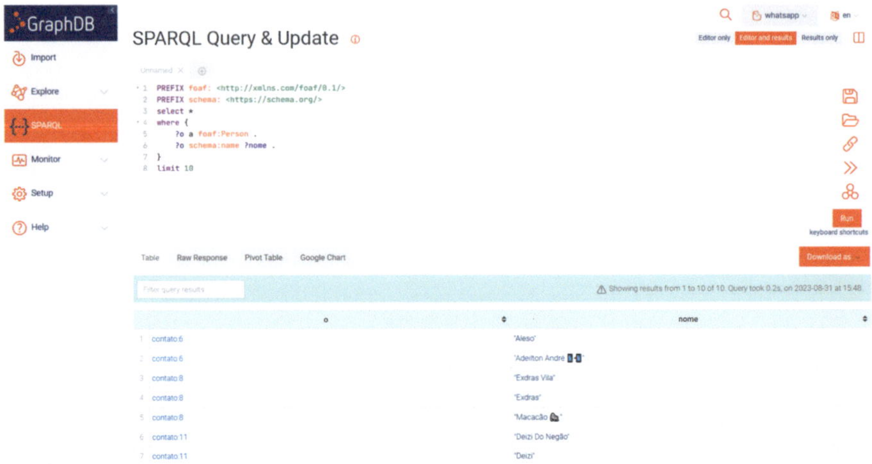

Fig. 6 SPARQL result after search for person with names

Table 2 Example of similarity correction

Term	Portuguese	English
Grasas a Deus	Graças a Deus	Thank God
tenho que aruma o dinheiro	tenho que arrumar o dinheiro	I have to get the money
Fas esse bagulho	Faz esse bagulho	Do this shit
Intelete do meu pai parol	Internet do meu pai parou	My father's internet is stopped
teresei	interessei	Interested

6 Case Study

Before using the presented code, the installation of dependency modules is required. Therefore, ensure that your environment has Python and the modules SPARQLWrapper, fuzzywuzzy, and NLTK installed.

After setting the parameters as described in the previous section, use the Python code to query the identified messages and apply the correction according to Table 2.

In Fig. 7, we can observe that despite using search keywords like "pino" (pin), "pedra" (rock), "camisa" (shirt), and "drogas" (drugs), we were able to retrieve "maconha" (marijuana), which is a type of drug, as well as a message where one interlocutor invites another to smoke.

The code used was made available via Pastebin[12] for better visualization.

[12] https://pastebin.com/XGc6js7X.

```
http://example.com/message/154537 - Tô com maconha aí em casa mn

http://example.com/message/154538 - Quando eu chegar vou aí pra nós fuma esse
```

Fig. 7 Partial result after run python script

7 Conclusions

This chapter explored the integration of text analysis into criminal investigation, highlighting the importance of innovative approaches that utilize the semantic web and ontologies. Although text analysis is a valuable tool, applying it manually can be time-consuming and prone to errors. The proposal to use ontologies in this chapter seeks to optimize this process, automating the analysis, and providing a more effective approach.

In the future, it is expected to create an application that expands the ability of law enforcement to solve cases and prevent crimes. The combination of advanced technology, text analytics, and ontologies offers significant potential to enhance criminal investigation and improve public safety more efficiently and effectively.

The proposal is promising for criminal investigation. It uses Semantic Web technologies and ontologies to facilitate text analysis and provide deeper insights into cases. A tool is still under development but has already been tested in some real cases. The initial results are promising, showing that the tool can identify patterns and hidden relationships.

Furthermore, the use of ontologies and the semantic web in criminal investigation offers an organizational and contextualized structure for the data. By integrating and standardizing heterogeneous information, these technologies facilitate the correlation between seemingly disconnected data, promoting a more comprehensive understanding of the investigative scenario and providing crucial support for strategic decision-making.

In summary, the joint application of text mining, ontologies, and semantic web in criminal investigation represents a significant advance in the ability of law enforcement professionals to deal with the increasing complexity of digital data. By integrating these technologies in an ethical and effective way, we can aspire to a future in which justice is more agile, precise, and equitable, guaranteeing the safety and protection of society.

References

Barreto, A. G., & Wendt, E. (2020). *Inteligência e investigação criminal em fontes abertas.* BRASPORT.

Behmer, E.-J., Chandramouli, K., Garrido, V., Mühlenberg, D., Müller, D., Müller, W., Pallmer, D., Pérez, F. J., Piatrik, T., & Vargas, C. (2019). Ontology population framework of MAGNETO for instantiating heterogeneous forensic data modalities. In J. MacIntyre, I. Maglogiannis,

L. Iliadis, & E. Pimenidis (Eds.), *Artificial Intelligence Applications and Innovations, IFIP Advances in Information and Communication Technology* (pp. 520–531). Springer International Publishing, Cham. https://doi.org/10.1007/978-3-030-19823-7_44

Brady, O., Overill, R., & Keppens, J. (2015). DESO: Addressing volume and variety in large-scale criminal cases. *Digital Investigation, 15*, 72–82. https://doi.org/10.1016/j.diin.2015.10.002

Carter, & David L. (2009). *Law enforcement intelligence: A guide for state, local, and tribal law enforcement agencies* (2nd ed.). Washington, D.C.: U.S. Department of Justice, Office of Community Oriented Policing Services.

Gonçalves, A. D. O., & Jacyntho, M. D. D. A. (2020). Um método para publicação semântica Linked Data de bases de dados convencionais e um estudo de caso real de artigos acadêmicos. *Transinformação, 32*, Article e180051. https://doi.org/10.1590/1678-9865202032e180051

Hendler, J., Allemang, D., Grobe, M. (2011). Semantic web for the working ontologist: effective modeling in RDFS and OWL (2nd ed.) Burlington: Morgan Kaufmann.

Manning, C., Raghavan, P., Schütze, H. (2008). *Introduction to information retrieval.* New York: Cambridge University Press.

Porter, A., Cunningham, S. (2005). *Tech mining: Exploiting new technologies for competitive advantage.* Hoboken: Wiley.

Investigating Tolerance for Uncertainty Through Text Mining: Enhancing Human Resource Strategy in Seed-Stage High-Tech Startups

Yushi Nakaya and Shuichi Ishida

1 Introduction

Background

Startup companies are poised to play a pivotal role in addressing both existing and emerging societal challenges, especially those known as "wicked problems," which are inherently complex and vexing (Stephen et al., 2021). Recent years have witnessed a surge in high-tech startups aiming to drive innovation with cutting-edge technology, setting themselves apart from businesses offering easily replicable solutions. The evolution of these startups, including changes in size and product status, occurs in parallel with their advancement through various funding rounds: the angel round, seed stage, series A, scale-up (series B and series C), and growth phases (series D and beyond). Notably, the seed stage emerges as particularly precarious and dynamic, marked by the business's establishment, identification of the ideal customer base, and the development of a Minimum Viable Product (MVP) (Failed, 2021; Tomczak & Brem, 2013; Zhang & Wang, 2017).

During the seed stage, the scope of individual work typically exceeds expectations found in larger corporations, a shift attributed mainly to a smaller workforce (Ycombinator, 2022). This phase also experiences significant fluctuations in organizational size, with personnel numbers often changing by a factor of 5 to 10 (Ycombinator, 2022).

Y. Nakaya (✉) · S. Ishida
Tohoku University, Sendai, Japan
e-mail: nakaya@lab.adansons.co.jp

S. Ishida
e-mail: shuichi.ishida.e6@tohoku.ac.jp

Y. Nakaya
Adansons Inc., Sendai, Japan

H. da Hora et al. (eds.), *Technology Mining*,
https://doi.org/10.1007/978-3-032-10849-4_10

Therefore, it is important for all members to work independently and think about what is missing in the organization on their own, and to take a leadership and ownership role in each of their own tasks. Thus, recruiting and training individuals for leadership roles and integrating them into the organization's core team become crucial. Such efforts are vital for shaping the company's future internal culture (Tannenbaum, 1966).

Managers of seed-stage startups, including the author, face the challenge of ensuring new hires align with the company's culture without risking disruption. Past experiences have shown that misalignments can lead to mutual consent terminations, underlining the severe and potentially irreversible consequences of poor hiring decisions on organizational culture. Addressing these risks is, therefore, of paramount importance.

An employee's high job performance, in other words, competency, is known to be related to his or her positive personality (Hu et al., 2021). In Hu et al.(2021), competency theory is described as follows.

The competency theory (McClelland, 1973) claimed that competency was the core factor for improving job performance: the higher the competency, the better the job performance. Luna-Arocas and Morley (2015) defined competency as "a set of observable dimensions, including individual knowledge, skills, attitudes, and behaviors, as well as collective team, process, and organizational capabilities, that are linked to high performance and provide the organization with sustainable competitive advantage."

Therefore, in seed stage startups, where the abilities and characteristics of employees have a significant impact on creating an organizational culture that is fundamental to the company's performance, it is extremely important to correctly assess and hire for competencies that are directly related to job performance, especially in order to lower the risk of poor hiring decisions, as mentioned earlier. However, in seed stage startups, there are a large number of things that founders need to do, such as product development, needs assessment, and conversations with investors, in addition to recruiting and training. Based on the author's experience, it is very difficult to find out whether or not a person has a high level of competencies that will enable him or her to perform well, so it is important to set hiring criteria and conduct hiring activities at a low cost.

While existing research largely focuses on methodologies vital for the growth of seed-stage startups and frameworks for harnessing and scaling technological assets (Atzmon et al., 2022), as well as decision-making post-COVID-19 (Parthasarathy, 2022), less attention has been given to the specific conditions and capabilities these organizations need. Although the significance of individual competencies in human resources, their organizational alignment, and their contribution to mitigating startup failure risks—the "valley of death" (Barr et al., 2009; Cummings et al., 2018; Markham et al., 2010; Thompson, 2018)—has been acknowledged, detailed exploration into these areas remains scarce.

Several studies have examined the relationship between individual capabilities in human resources and aspects such as organizational culture, leadership, and motivation within larger business contexts (Schein, 2016; Tannenbaum, 1966). However, these investigations have predominantly focused on larger, publicly traded companies, with startups in the seed stage receiving limited scrutiny. Although some surveys have included small and seed-stage startups, their quantitative analyses usually encompass more mature companies, restricted to scale and growth stages (Men, 2021). Moreover, despite some research on entrepreneurship's link to leadership absence (Girdauskiene et al., 2015), there is a lack of studies differentiating startups by their growth stages and how specific deficiencies affect them. A notable reference (Gbadegeshin et al., 2022) aggregates research on risk factors leading to startup failure without distinguishing among the startup phases.

Consequently, there is an urgent need for a unified framework capable of systematically assessing and identifying the competencies related to leadership and organizational core strengths during the seed stage of high-tech startups.

Prior research has shown that uncertain events are common in seed-stage startups.

Startups attempting to innovate through more advanced technologies face higher uncertainty levels (Borracci et al., 2021; Byrne et al., 2020; Felsman et al., 2020), contributing to stress and rapid organizational changes (Tannenbaum, 1966). Therefore, in a startup with a small number of employees in the seed phase, employees as well as the Founder are likely to be exposed to a high level of uncertainty, and in order to achieve high performance in the face of such uncertainty, tolerance to these rapid changes and uncertain events is considered to be related to competencies.

Purpose

Recruiting initial team members is crucial, as they significantly influence the company's culture. During the seed phase, management often grapples with resource constraints, impacting training, personnel evaluation, and critical hiring and termination decisions. Additionally, there is a pressing need for an effective method to gauge potential recruits' alignment with the company's culture and characteristics, and their ability to take on leadership roles and form the organization's nucleus. The author also hypothesized that startup employees are exposed to high uncertainty during the seed phase as mentioned earlier, and that their tolerance for uncertainty has a lot to do with their competencies in order to achieve high performance in this environment.

Thus, this study poses a research question: Are the abilities of tolerance for uncertainty crucial for individuals to form the core of an organization, inspire fellow team members, and contribute substantial value?

Our primary goal is to develop a framework that streamlines the identification of these essential competencies. This framework will facilitate their incorporation into human resource strategies, including recruitment and training, specifically designed for high-tech startups in the seed stage.

The ultimate objective of our research is to create a methodology that enables seed-stage startups to make informed hiring and firing decisions more efficiently. This is particularly vital for organizations with limited resources for these processes but are keen on expanding their organizational scale through strategic recruitment.

2 Prior Research

Tolerance for Uncertainty

Many studies have used the concept of human uncertainty tolerance (Dantan et al., 2013; Lawrence & Lorsch, 1967; Lubart, 2016; Thompson, 1967).

The study in Lubart (2016) investigated uncertainty tolerance, personality traits, and their relationship to specialty selection among medical students considering various aspects of their careers (risk, ambiguity, complexity). Students with high uncertainty at the beginning of their studies show reduced tolerance over time, while those with low tolerance tend to increase it after enrollment. However, the cause of these changes remains unclear and further interventions may have affected the results.

On the other hand, studies that took a psychological approach include (Dantan et al., 2013) and (Thompson, 1967), which studied the influencing factors.

The research in Dantan et al. (2013) posits that the ability to think divergently, the polar opposite of convergent thinking observed among patients suffering from mental health disorders such as depression, is closely linked to uncertainty tolerance. The appropriateness of interventions such as improvisational theater as a therapeutic method for individuals unable to think of alternatives in response to a single event was investigated. Divergent thinking has been associated with success in creative fields such as entrepreneurship (Lawrence & Lorsch, 1967). Hence, in Thompson (1967), the authors medically examined the effects of changes in saliva volume on the ability to make decisions under uncertainty and risk and found that acute stress improved decision-making skills and increased tolerance levels.

Although methodological studies have used uncertainty as a probability distribution (Knight, 1921), creating a probability distribution model for each human statement or situation is challenging, and no such previous studies exist.

Studies have shown that the external environment of management organizations is full of complexity and uncertainty (Bhattacharya et al., 2017; Iyke, 2020).

Based on these previous studies, uncertainty tolerance in seed stage startups may be associated with the ability to think divergently while improving decision-making.

Because these are related to entrepreneurial success, uncertainty tolerance may also be important in the seed stage high-tech startups, which are the target of this study.

In addition, as discussed in Sect. 1.1, prior studies such as (Borracci et al., 2021; Byrne et al., 2020; Felsman et al., 2020) have shown that uncertain events are common in seed-stage startups, and employees as well as founders are likely to be exposed to high uncertainty. Therefore, these characteristics are very important for seed-stage startups where the overall performance of the employees is likely to be directly related to the company's performance.

Method of Analyzing Human Resource Capabilities in a Company

Previous studies have used text mining to analyze the connection between motivation, leadership, and company results, using meeting minutes, daily reports, and other data (Blair et al., 2014). The method used in these studies is to conduct text mining first, followed by quantitative analysis, and finally, qualitative evaluation of the results. Some studies using text mining have been conducted on startups (Bzhalava et al., 2018; Ilya et al., 2018; Sung-Soo & Yang, 2017), but most are limited to the analysis of startup profiles and portfolios, whereas some use text mining to delve deeper into the characteristics of individual organizations, such as large firms.

It is desirable to generalize the conditions of these studies and create a standard framework using an actual analysis rather than through case studies. The population is assumed to be similar, even if the dates and times of meetings differ. Therefore, we will use the method in (Ishida & Kin, 2012) and other methods to apply the parameters obtained from the quantitative analysis to other data groups.

3 Methodology

Research Question

For small organizations, it is essential to be involved with the external environment while making managerial and recruitment decisions. The organization will be structured considering various uncertainties, including the company's survival.

In (Dantan et al., 2013; Lubart, 2016), and others, the ability to think divergently is considered as the ability to tolerate uncertainty and high risk. Moreover, in Lubart (2016), it is shown that this ability leads to entrepreneurial success.

In this study, the authors hypothesize that the ability to tolerate uncertainty is related to the research question of what the necessary human resource strategies of a high-tech startup organization in the seed stage are. Therefore, the research question for this study is that uncertainty tolerance is deeply linked to competence in startups. The hypothesis for quantitatively examining uncertainty tolerance in this research question is based on the previous literature (Thompson, 1967), which suggests that uncertainty tolerance can be examined by examining whether conversations are not too focused on a single topic or a single word, and whether they are moderately divergent. Therefore, the authors hypothesized that examining the divergent nature of conversations during interviews could lead to an assessment of uncertainty tolerance and competency in startups.

Planning the Analysis Objectives

We noted in the previous section that uncertainty tolerance as a hypothesis of competency in seed stage startups may be manifested in whether or not they converge on a single topic or word in a conversation.

To test this hypothesis, we thought we could examine some of the differences in uncertainty tolerance by analyzing the differences between employees who continue to perform well in seed-stage startups and those who leave because they do not perform well.

Since uncertainty abounds on a daily basis, we can examine uncertainty tolerance by examining the text of how people speak when uncertain events occur in their daily conversations. Here, recordings of meetings at Adansons Inc, which the author manages and conducts action research for, will be used for analysis.

When selecting targets for analysis, special consideration will be given to the objective of comprehending how day-to-day operations tolerate uncertainty and the outcomes that ensue. Recordings of online meeting tools are selected and used for analysis at the most appropriate meetings during the action research. Adansons Inc. possesses several patents for technologies that serve as the foundation for Tohoku University's innovations. These technologies are exceptionally advanced, boasting a multitude of potential applications that necessitate corresponding theoretical extensions. Additionally, they encompass elements and potentials that remain undetermined. From a management perspective, this technology stack ranks among the most uncertain. Therefore, Adansons Inc. has a committee to understand these technologies, find the correct usage of each technology in different areas, search for technologies that are currently lacking, and discuss the possibility of their business applications, including formulating a direction, sharing results and knowledge, and implementing in-house demonstration for various use cases. Uncertainties in the technology and market must be considered to determine whether a technology is inapplicable in a particular area or can be applied through technological expansion. The content of the statements includes decisions about considering uncertainties in formulating the company's next course of action. Furthermore, this committee had members who participated only in January 2021–September 2021 and those who subsequently left the company due to disagreements with the organizational culture. We analyze the differences in characteristics between members central to the organization, other members, and members who have left the organization from the perspective of what kind of differences in speech characteristics led to a sense of disagreement with the organizational culture.

There were five participants in this committee; two managers and three engineers, one of the engineers being an employee who had left the company due to disagreements in thinking and culture. The characteristics of the five participants, referred to as Subjects 1–5, are as follows:

Subject 1: Employee, engineer. From 2020 to the present, he has been highly central in the organization and is in a position similar to that of a manager. Although he has theoretical knowledge, he actively studies what he does not know and tries to make the most of it. The author's experience before the analysis is that Subject 1 can formulate his own hypotheses and opinions, not limited to this meeting group, on unknown events that are often unknown even to the author and his management team.

Subject 2: Former employee, now retired. Engineer. He has felt friction since the committee was formed due to discrepancies in the direction of conversations and ideas. Although unfamiliar with methods or patents, his mathematical and programming skills are the best in the group. In technical validation, he would often independently work with the content, performing tasks outside what he had been instructed, but sometimes went off the rails due to differences in thinking.

Subject 3: An engineer and tech lead who is still active. Like Subject 2, he has good mathematical and programming skills and is most familiar with methods and theories related to patents. Moreover, as a Tech Lead, he is involved in developing new theories and technologies within the organization. The author's experience prior to the analysis is that Subject 3 can see the big picture, which allows him to take conversations smoothly. In addition, because of his ability to see the whole picture, he stays calm and within the general framework, even when doing something outside the instructions during the technical verification process.

Subject 4: The board member, author, and CTO who moderated this meeting. As the CTO, he has a deep understanding of patented technologies. However, his role is to control the overall direction of the meeting so that the discussion stays on-topic; hence, his methodological comments are less frequent than those of Subject 5.

Subject 5: Executive, product manager. In this committee, he is central in explaining the patented technology, discussing aspects other members may overlook, and providing direction to the employees, including technical comments on programming. He has the deepest understanding of patented technologies in the company and is well-versed in the theories necessary to extend the technology.

The meetings occurred on a weekly basis, with occasional absences by one of the five members. All meetings were recorded through both video and audio using an online conferencing tool.

The focus of the meetings, which forms the basis for our analysis, encompassed discussions where two members of the management team, possessing expertise in patented technologies, imparted knowledge to other team members on how to utilize these technologies effectively. Additionally, each member explored strategies for applying patented and related technologies to specific use cases with the aim of enhancing their effectiveness and practicality. The meetings were structured around the objective of instructing employees in how to approach the utilization of patented technologies, as well as how to analyze and evaluate data within this context. While the two management team members possessed the knowledge of how to navigate such scenarios, there were numerous aspects open to discussion, primarily due to the inherent uncertainty associated with these endeavors. Moreover, discussions focused on interpreting the results obtained through the use of patented technology, acknowledging that such outcomes can often be subject to multiple interpretations. Consequently, this meeting group was selected as the subject of analysis, as it aligns well with the research question's objective, fostering discussions amidst a backdrop of uncertainty. Additionally, the weekly meeting cadence imposed a constant need to formulate actions for the subsequent week, even in the face of uncertainty.

It is important to note that there were only five participants and only one retired subject. Therefore, it is possible that there may be a tendency or bias specific to that one person. However, this study will be reported as a case study, examining the relationship between the characteristics and competencies possessed by the person who retired as a matter of fact.

In addition, my personal feelings and biases should be eliminated as much as possible. Therefore, quantitative analysis methods such as PCA were used to determine whether retired subjects had different tendencies on a numerical basis, so that the results of the study would not differ due to biases beyond necessity.

Text Mining

The premise of this study is that it is a case study of a single company, and by generalizing it, we aim to make it applicable to other companies. Therefore, the analysis will be qualitative and not quantitative. However, generalizations will be made regarding possible conditions and environments so that the analysis is not limited to the case study category.

As an analysis method, we visualize the meeting content, including spoken words, and convert the recorded data from the meeting into a form that can be analyzed by text mining to generalize the object of analysis. The use of text mining in such analyses is common, and numerous similar prior studies exist (Bzhalava et al., 2018; Ilya et al., 2018; Sung-Soo & Yang, 2017).

When performing text mining, the data is separated into words and phrases, the frequency of occurrence of each word is calculated, and the list of extracted words is created as a dataset. KHCoder (Corley et al., 2007) is used to create the list; the number of occurrences is recorded after morphological analysis and segmentation.

During word extraction, it is common to narrow down to a certain part of speech or exclude exclamations, especially when looking for common characteristics. However, words related to stilted speech, such as "ah" and "e," are strongly remembered, and although not meaningful by themselves, they make it difficult to predict the word immediately following them. Moreover, fluent and non-fluent speakers differ in the ease with which they convey their conversations, including parts of sentences that make others strongly aware of them (Hlavac, 2011).

In contrast, slowness and sentence breaks are important in clarifying the speaker's feelings (Clark et al., 2002) and can lead to expressing intentions such as the desire to transfer the right to speak to another (Ewing et al. 2017).

In this study, we assumed that it is unclear what characteristics of uncertainty tolerance are expressed in words and what trivial behaviors other than the purpose of conversation may be involved in such characteristics. Hence, we treated all extracted words as equally important and analyzed them without restrictions in terms of parts of speech.

Quantitative Analysis

Quantitative analysis was performed on the list of extracted words. In this study, the analysis focused on the behavior of retired Subject 2 compared to that of the other subjects.

As a method for quantitative analysis using text mining, a co-occurrence network was first used to visualize the frequency of occurrence of words and the relationship between subjects, and then a qualitative evaluation is conducted. A co-occurrence network is a diagrammatic representation of commonalities between words in a sentence and is a well-known text mining technique. To create the co-occurrence network, the Jaccard coefficient calculates similarity in texts based on the frequency with which two words appear simultaneously; the higher the coefficient, the higher the similarity. The Jaccard coefficient is a measure of the similarity of sets, and in text mining, it is a measure of the similarity, or distance, between sentences. Weights are then calculated considering which words appear more frequently with which speakers; the higher the weight, the thicker the line between speakers and words in the plotted co-occurrence network. Thus, by visualizing the co-occurrence network, it is possible to observe objectively which speaker utters which words more often. Moreover, co-occurrence networks assume that the speaking habits of each person are also meaningful in expressing the characteristics and personality of each person, and analysis is performed including all parts of speech that can be included in KHCoder's morphological analysis, such as adverbs, emotional verbs, and suffixes. The top 100 words were used to draw co-occurrence relations.

After visualizing the co-occurrence network, a quantitative analysis method examines the words uttered and speaker characteristics.

Quantitative text mining analysis includes correspondence analysis and principal component analysis (PCA) (Greenacre, 2017). Correspondence analysis is widely used as it can easily visualize the relationship between categories (Ihaka & Gentleman, 1996). However, it does not consider the number of word samples, making it difficult to measure which words contribute significantly to separating specific speakers or identifying characteristics. In contrast, PCA considers the number of samples and analyzes the relationship between each principal component and the speaker using contribution analysis to each principal component. PCA searches for specific cohesion in cluster analysis when the number of samples is sufficiently large and quantitative analysis is possible. However, in this study, a sample size based on the number of utterances per meeting can be considered a sufficient sample size. Therefore, PCA, which is more advanced than correspondence analysis, was adapted to qualitatively analyze the characteristics of the entire meeting based on the overall results.

PCA was performed using Scikit-learn (Rossum & Drake, 2009), a library in Python (Pedregosa et al., 2011), which runs PCA on a tabular dataset that records the number of words uttered by each subject on the vertical and horizontal axes. The values were standardized by subtracting the mean of all utterances from the number of times which speaker uttered which word and dividing it by the standard deviation so that the mean was 0 and the standard deviation was 1. The reason for using standardization rather than normalization is that the upper limit of utterance for each word at each meeting session is different. Similar treatment of various parts of speech may result in large differences for each part and consequently bias the results for one part of speech. PCA provides a vector of scores for each principal component for each speaker, displayed simultaneously with the vector of each word on a vector space based on each principal component, to visualize which speakers have what word component strongly, even when clustering by word is difficult. The relationship between the principal components and the corresponding score vectors was then examined to investigate what characteristics differentiated Subject 2 from the other subjects.

Also, when performing PCA, it is expected that there will be multiple principal components. Since this is a meeting where all participants are discussing a common agenda, there is a possibility that words related to the agenda that make up the majority of the conversation may appear specifically in the first principal component, etc. If a global principal component appears that has a contribution rate of over 70%, it shall be excluded from the analysis and the remaining components shall be analyzed. The principal components used in the analysis shall be selected according to the number of participants in the meeting being analyzed. For example, if there are five participants, up to the fifth principal component shall be analyzed, and if there are four participants, up to the fourth principal component shall be analyzed. However, the number of principal components for which information related to the subject being tracked will vary from one analysis data to another and cannot be determined in advance. Therefore, we will utilize principal component vectors to calculate correlations to determine which subject's component is dominant in which principal component each time.

Dataset

In this study, we used "Whisper," a large language model created by OpenAI, to perform highly accurate automatic transcription (Radford et al., 2022). In recent years, deep learning-based large language models have become a mainstream transcription method. Whisper was selected for this analysis because the pre-trained model was free and easy to use, and the documentation on how to use it is publicly available and widely recognized by hightech venture companies. However, speaker annotation and transcription error corrections were made manually.

For the analysis, three meetings with distinctive features were chosen: The first meeting took place on July 22, 2021. During this session, the executive provided a unilateral explanation to the employee regarding the meeting's objectives and the rationale behind task assignments. This meeting also included a brief question and answer session. The second meeting, held on August 19, 2021 involved progress reports from each employee (excluding the executive) and instructions from the executive regarding necessary actions and findings.

The third dataset was derived from a meeting that took place on August 26, 2021. Similar to the second dataset, this meeting involved employees presenting progress reports, followed by mutual feedback, including input from the executives. Following this, employees expressed their opinions regarding necessary actions. The author's selection comprised three types of meetings: those in which instructions were the primary focus, meetings in which employees both reported their opinions and provided instructions, and meetings where the emphasis was placed on employees' opinions.

One caveat to this dataset is that there is only one retiree, and it is possible that characteristics specific to that retiree will be perceived as distant from competency. However, if the person who actually retired has characteristics that are different from others, it is necessary to identify those specific characteristics in this case study. On the other hand, considering the possibility that the author has a bias that the retired person has characteristics different from others that are idiosyncratic, we have tried to eliminate that bias and enable a neutral evaluation by combining quantitative methods as well as qualitative evaluation.

4 Analysis

Word Extraction Results by Morphological Analysis

The results of the occurrence counts are shown after morphological analysis and sharpened processing were performed using KHCoder. The analysis was performed for each of the three datasets.

The extraction results showed that the first dataset had 217 sentences and 129 utterances, containing 10,090 words, with 1228 different words. The second dataset had 259 sentences uttered 121 times, of which 11,258 words were extracted, with 1259 different words. The third dataset had 262 sentences and 121 utterances, containing 7457 words, with 972 different words.

Co-occurrence Network

Figure 1 shows the results of displaying the co-occurrence network for the first meeting. The co-occurrence network provides a visual display of the relationship between words and speakers. However, to convert this qualitative study to a quantitative study and eliminate observer bias, PCA, described below, should be used.

A tendency to use more technical terms is observed in Subjects 1 and 2. In addition, all but Subject 2 use the word "do" often.

Figure 2 displays the co-occurring network for the third dataset. Notably, Subject 2 frequently referenced specific names of analysis methods, whereas Subjects 1, 4, and 5 employed words suggesting vagueness, such as "something," more frequently. Observations from the co-occurrence networks across all three meetings reveal that Subjects 2, 3, and 5 frequently employ technical terminology, including method names unrelated to patented technology. In contrast, Subjects 1 and 4 utilize words beyond method names and exhibit connections to all other subjects through these words.

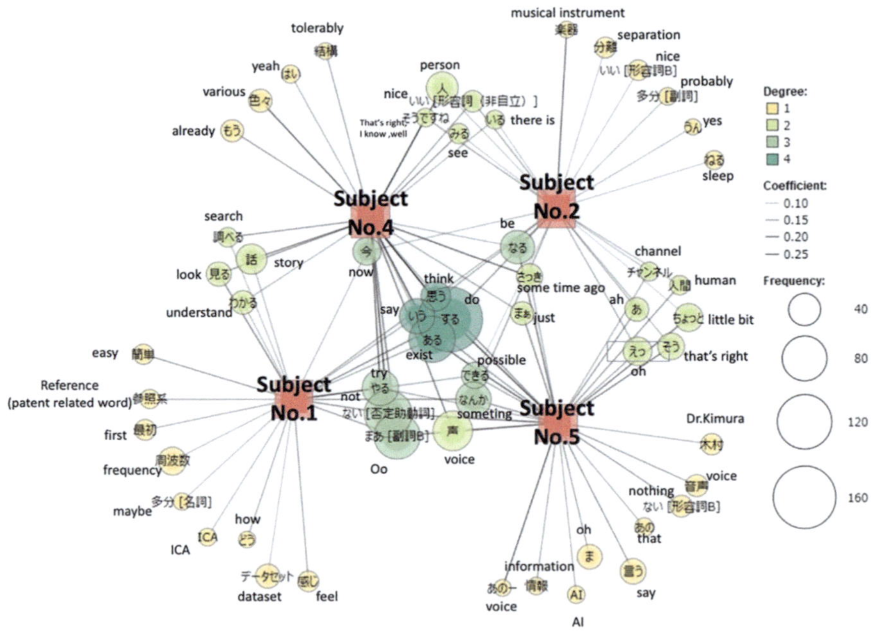

Fig. 1 Co-occurrence network for the July 22, 2021meeting

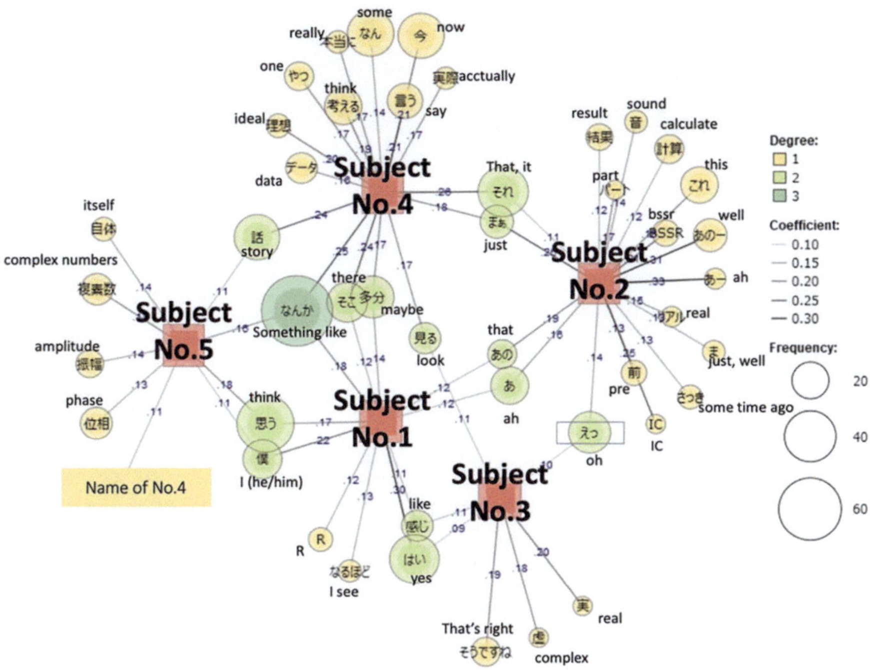

Fig. 2 Co-occurrence network for the August 26, 2021 meeting

Principal Component Analysis

We begin by providing the results of the PCA conducted for the initial meeting. A table was created, featuring subject names along the vertical axis and words along the horizontal axis, as illustrated in Fig. 3. Following the construction of this matrix, PCA analysis was performed.

The principal components were decomposed into four groups, corresponding to the number of participants. First, Fig. 4 displays the principal component scores, where each principal component's name is depicted on the horizontal axis, and subject numbers are presented on the vertical axis. The vertical axis quantifies the contribution of each principal component when expressed in terms of the four principal components, facilitating the identification of patterns and associations within each. Additionally, PC1 and PC2 are indicated sequentially, starting with the first principal component.

The calculated contribution ratio of each principal component and cumulative contribution ratio are shown in Figs. 5 and 6, respectively.

	talk	somethin g	easy	What?	Referenc e (patent related word)	area	me	various	...
Subject4	1.477098	-1.214286	0.055556	-0.648901	-0.405751	0.150756	-0.160817	1.459993	...
Subject2	-0.738549	-0.357143	-0.611111	0.034153	-0.676252	-0.753778	-0.804084	-0.811107	...
Subject1	-0.369274	1.071429	1.388889	-0.785512	1.487755	1.356801	1.447352	-0.324443	...
Subject5	-0.369274	0.5	-0.833333	1.400261	-0.405751	-0.753778	-0.482451	-0.324443	...

Fig. 3 Part of the data standardized for the number of utterances at the meeting on July 22, 2021

	PC1	PC2	PC3	PC4
Subject4	-2.846651	5.528277	-2.01736	-1.33E-15
Subject2	-4.677224	-2.040037	4.103563	4.09E-15
Subject1	0.163256	-4.428146	-4.002342	-7.45E-15
Subject5	7.360619	0.939906	1.916139	4.69E-15

Fig. 4 Principal component scores when principal component analysis (PCA) is applied to the July 22, 2021, meeting

PC1	4.68E-01
PC2	3.07E-01
PC3	2.26E-01
PC4	2.71E-33

Fig. 5 Contribution of PCA when applied to the July 22, 2021, meeting

Almost 100% of the information is explained by up to PC3. A portion of each principal component vector calculated using Scikit-learn's pca.components is shown in Fig. A1 of the Appendix Section. The vertical and horizontal axes represent each principal component name and word, respectively.

To analyze the relationship between each principal component and subject, we calculated the correlation coefficient between the matrix showing the number of utterances of each word for each subject (Fig. 3) and the principal component vector matrix showing the relationship between each principal component and word calculated using PCA (Fig. 7). Correlation coefficient is an index that measures the strength of a linear relationship between two sets of data and takes a value between -1 and 1

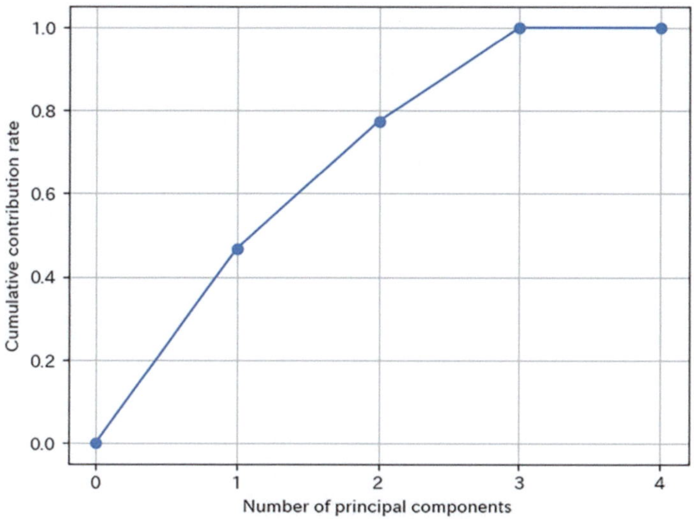

Fig. 6 Cumulative contribution of PCA applied to the July 22, 2021, meeting

as a real number. 0 is uncorrelated. And the larger the value, the stronger the positive correlation, i.e., the higher the relationship. The results in Fig. A2 of the Appendix Section show the color map created based on the maximum and minimum values of the correlation coefficients. The larger the value, the higher the correlation coefficient, and the greater the positive correlation, the darker the color. A correlation coefficient between 0.7 and 1 is considered a strong positive correlation (Ratner, 2009).

Figure A3 of the Appendix Section shows the components of each principal component vector, including PC1 having a particularly high positive correlation only for Subject 2, and words that are positive and exceeding the thresholded value of 0.1 for each principal component (Figs. 7, 8 and 9).

Next, we show the relationship between words and speakers when using correspondence analysis. However, due to the small number of samples and the large variance, it is difficult to perform clustering. Therefore, the principal component score vector for each speaker obtained using PCA is displayed simultaneously with the vector of each word on a vector space based on each principal component to visualize the relationship between each speaker and word, as well as distribution bias. Figures 10, 11 and 12 show the relationship between each principal component and word in a scatterplot.

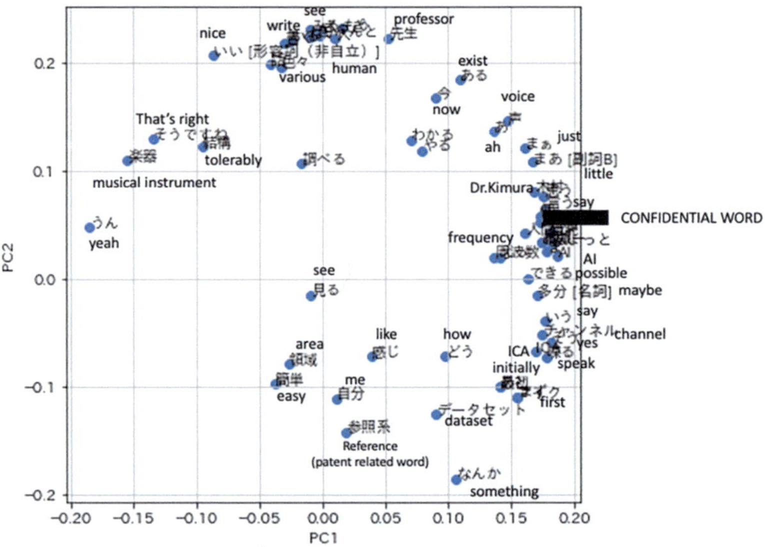

Fig. 7 Mapping of each word on the space defined by PC1 and PC2 when PCA was conducted for the July 22, 2021, meeting

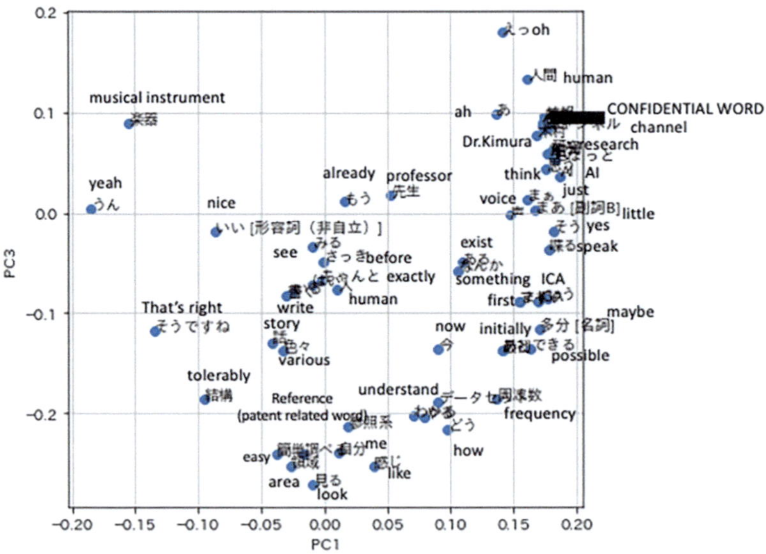

Fig. 8 Mapping of each word on the space defined by PC1 and PC3 when PCA was conducted for the July 22, 2021, meeting

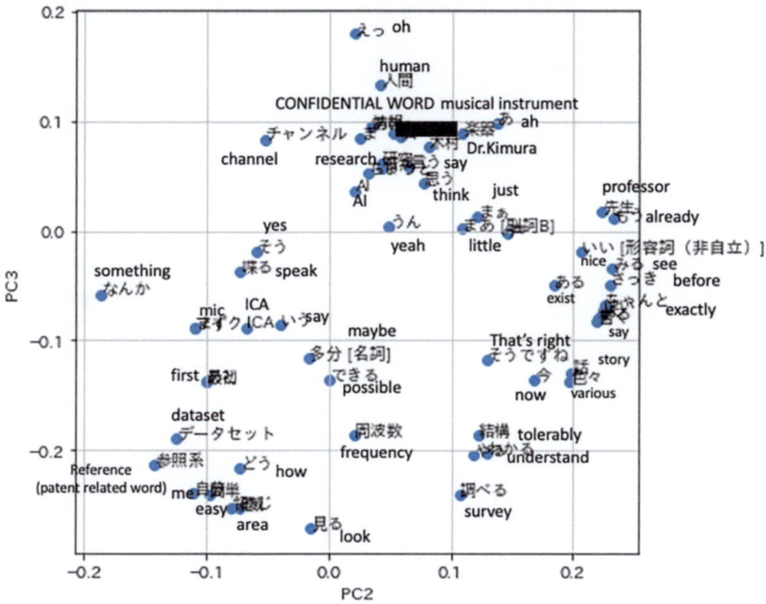

Fig. 9 Mapping each word on the space defined by PC2 and PC3 when PCA was conducted for the July 22, 2021, meeting

Next, the principal component score vectors for each speaker obtained using PCA are visualized by displaying them simultaneously with the vector of each word on a vector space based on each principal component, as shown in Figs. 13, 14 and 15. Principal component scores are the values the data take on each principal component axis.

If the number of participants is large and difficult to separate, the analysis can be extended to multiple dimensions, as shown in Fig. 16.

These procedures were followed for the two meetings in August. Below are excerpts of the characteristic plots from each meeting. Figures 14 and 15 show the mapping of each word on the space defined by PC2 and PC3 for the August 19, 2021 meeting and the plot of principal component vectors. Figures 16 and 17 show the mapping of each word onto the space defined by PC2 and PC4 for the August 26, 2021 meeting and the plot of the principal component vectors.

The analysis results for all meetings showed that Subject 2 was characteristically isolated, with a strong correlation to technical terms.

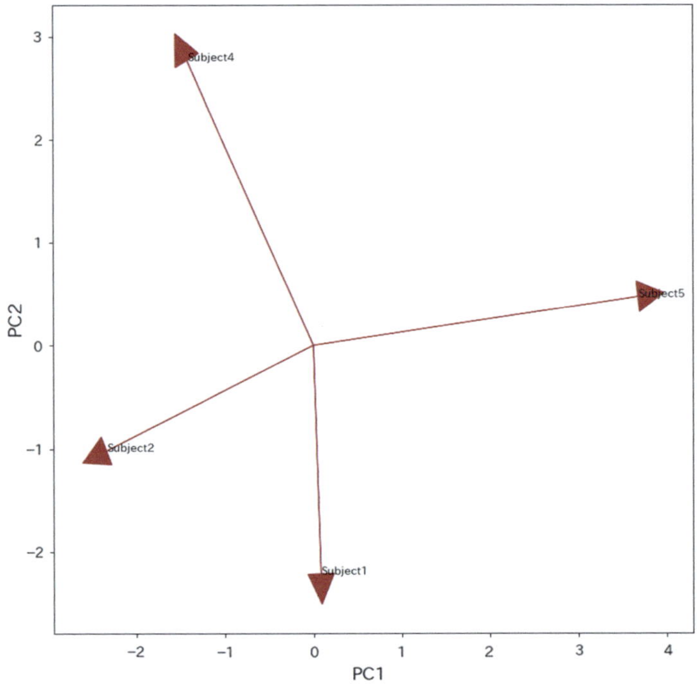

Fig. 10 Principal component vectors for each speaker on the space defined by PC1 and PC2 when PCA was conducted for the July 22, 2021 meeting

5 Discussion

Features from Quantitative Analysis

The PCA results reveal that Subject 2 exhibits distinct components compared to the other subjects. In this section, we delve into the interpretation of these components derived from each analysis.

Initially, it is observed that the components strongly correlated with Subject 2 are either uncorrelated or exhibit a pronounced negative correlation with the other subjects. Consequently, the relevant principal components offer the capability to distinguish Subject 2 from the rest. As such, we will conduct a detailed examination of the words closely associated with Subject 2 and their characteristics. This assessment will consider the frequency of words that might not have been adequately addressed in the correspondence analysis.

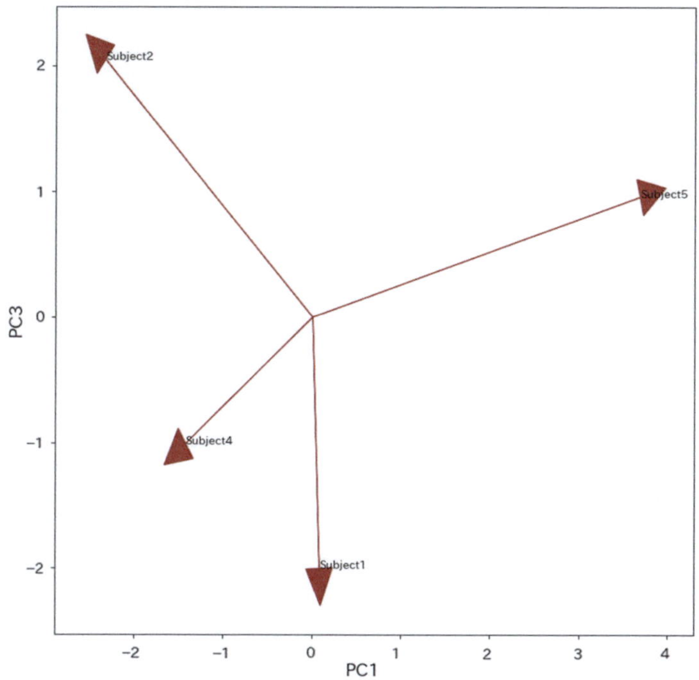

Fig. 11 Principal component vectors for each speaker on the space defined by PC1 and PC3 when PCA was conducted for the July 22, 2021 meeting

In the case of the meeting held on July 22, 2021, PC1 displays a high correlation coefficient of 0.85 with Subject 2, with a notable emphasis on terms related to patents and technical jargon. Technical terms are distributed across PC1 to PC3, with PC1 encompassing words used by Subject 4 to elucidate concepts to other participants, PC3 containing technical terms associated with patents, and PC2 encompassing other technical terminology. Furthermore, during the August 26, 2021 meeting, Subject 2 exhibits a strong correlation with PC4, with a coefficient of 0.97. This is particularly conspicuous in the case of technical terms unrelated to patents.

These findings suggest that Subject 2 employs a more extensive array of technical terminology compared to the other subjects, and these characteristics are closely associated with the use of technical terms. Furthermore, it is notable that technical terms, even those not directly linked to patented technology or the central focus of the conversation, may occur more frequently in Subject 2's contributions.

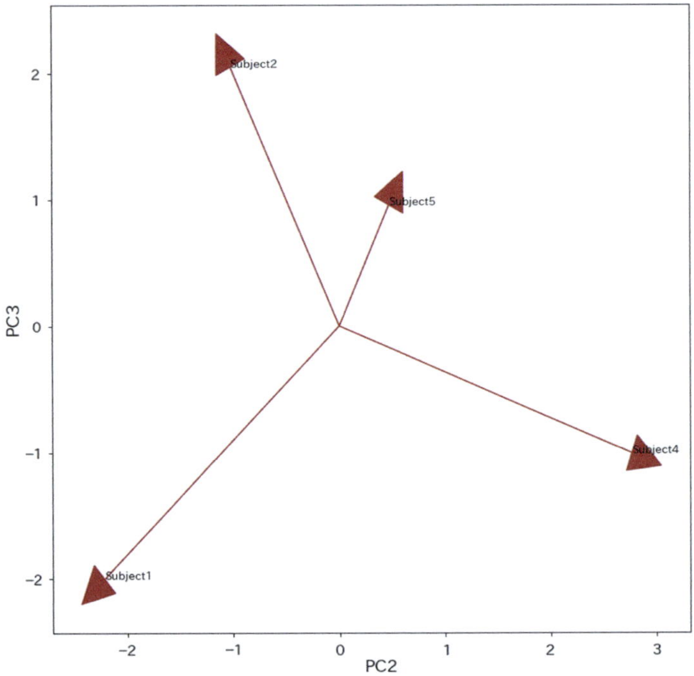

Fig. 12 Principal component vectors for each speaker on the space defined by PC2 and PC3 when PCA was conducted for the July 22, 2021 meeting

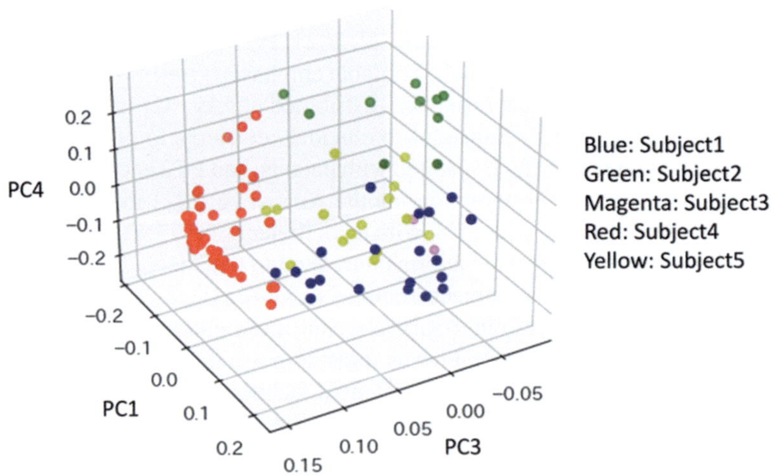

Fig. 13 Scatter plots for PC1, PC3, and PC4 based on principal component analysis

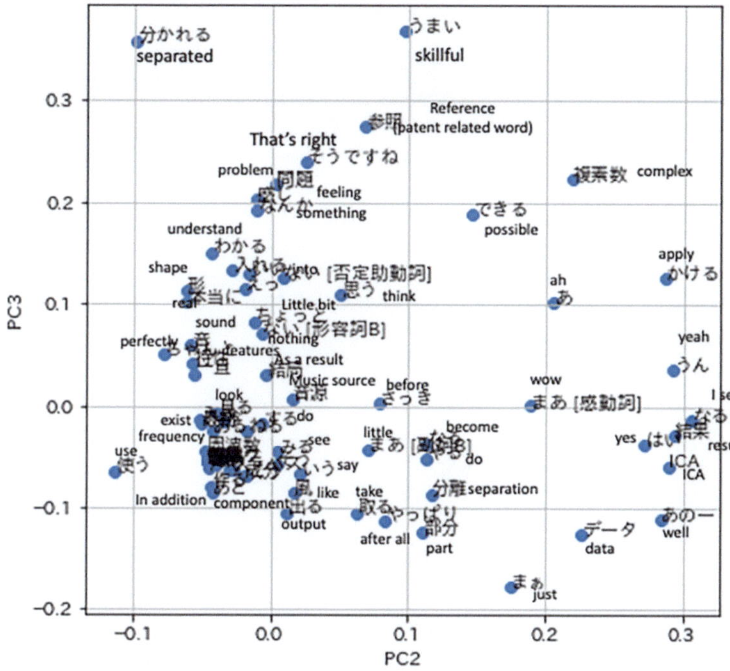

Fig. 14 Mapping of each word on the space defined by PC2 and PC3 when PCA was conducted for the August 19, 2021 meeting

Relationship to Uncertainty Tolerance

Prior research has shown that uncertainty tolerance is largely related to divergent thinking and does not converge in response to uncertainty and risk types (Lubart, 2016; Matta et al., 2022). However, there is a difference between Subject 2 and other subjects in terms of whether there were more words with technical terms for the patented technology whose usage was non-trivial. In addition, discussions using technical words not directly encompassed by the patented technology were particularly noticeable. The qualitative evaluation reveals that Subject 2 engaged in discussions on methods to verify the patented technology, unlike the purpose of the meeting, which was to understand how to use the patented technology and explore its application areas.

The nature of the discussion may be considered similar to convergent thinking in prior studies. Thus, more discussions using technical terms than business-like or schematic discussions are possible.

In addition, such discussions of technical terms occur not in progress reports and the like but rather when formulating what to do next, the so-called Next Action. The ability to think about the next actions during uncertainty and the rate at which methodological conversations are uttered may have contributed to this. Therefore,

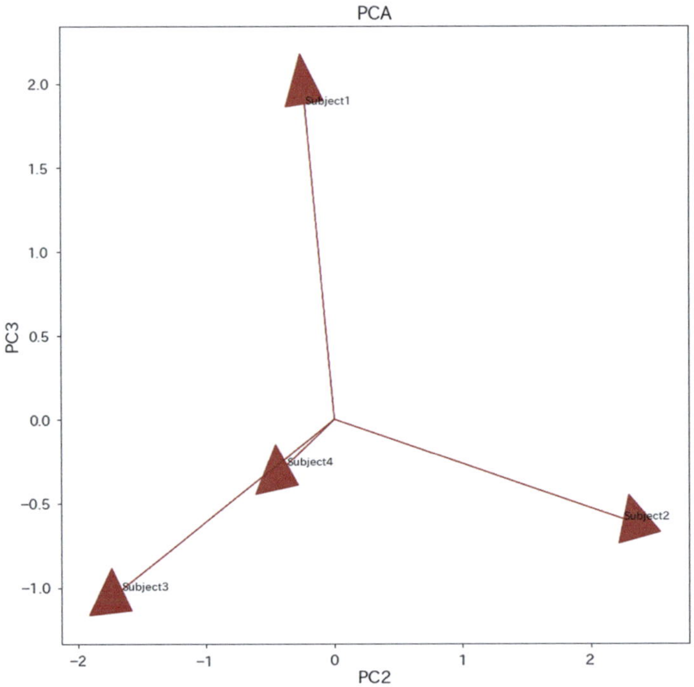

Fig. 15 Principal component vectors for each speaker on the space defined PC2 and PC3 when PCA was conducted for the August 19, 2021 meeting

the ability to tolerate uncertain content means the ability to make assumptions as well as facts in the face of uncertainty and be aware of the fundamental premise without being bound by the methodology.

Challenges in Utilizing the Method as a Framework

This section explores the applicability of the findings from this study to other meetings and organizations, along with associated challenges.

Using the components derived from PCA, it is possible to project these components by taking the dot product of eigenvectors from the variance–covariance matrix, as exemplified by the August 26 meeting, and comparing them with those obtained through PCA in similar analyses of other meetings. Variance–covariance matrix is a matrix representation of the covariance between the elements of a vector. It is an extension of the concept of variance in a scalar-valued random variable to multiple dimensions. This allows for the superimposition of principal component vectors. However, it is important to ensure consistency in the words used across meetings. Even when statements across multiple meetings share similarities in word usage,

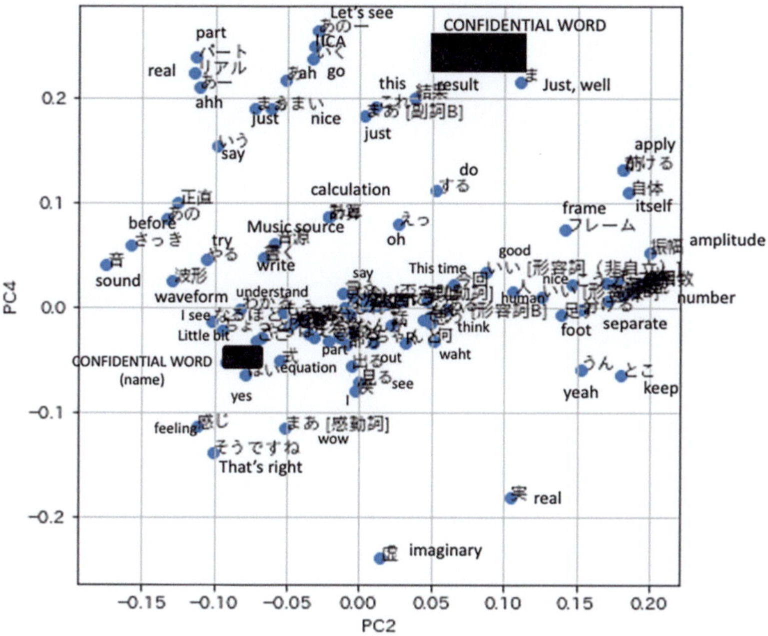

Fig. 16 Mapping of each word within the space defined by PC2 and PC4 when PCA was conducted for the August 26, 2021 meeting

there may still be differences, complicating the universal application of eigenvectors from the variance–covariance matrix in all cases.

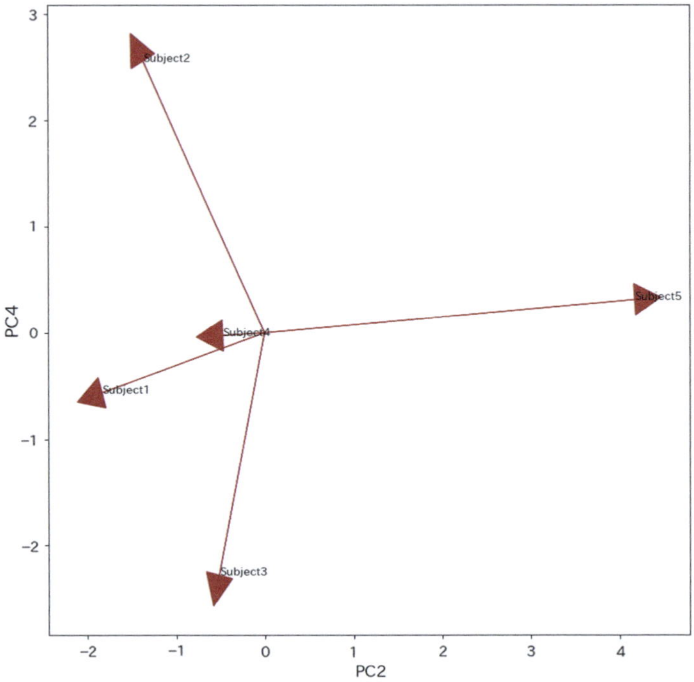

Fig. 17 Principal component vectors for each speaker within the space defined by PC2 and PC4 during PCA conducted for the August 26, 2021 meeting

The process of transcribing the three meetings, text mining to create a matrix of speakers and words, and using PCA to separate them is expected to be used as a framework. The separation of Subject 2 is reproducible in terms of visualizing using correlation coefficients, separating using a principal component vector and two principal components, and identifying components strongly related to the subject. The framework can detect differences between subjects by identifying components and analyzing words in the identified components. In addition, the framework could analyze speech tendencies over time, which could be transferred to an analysis of what kind of people may change due to their relationships with other people.

However, given the relatively small number of subjects compared to the data quadrants, it is challenging to rely solely on the visualization of principal component vectors for accurate separation. To establish a broadly applicable framework, expanding the dataset with more participants and meetings and confirming reproducibility under these conditions is essential.

Framework Applicability of Different Stages Startups and Other Industries

In Sect. 5.3, we discussed the versatility of the matrix calculation method to discover who tends to differ from others in their tolerance for uncertainty technically as a framework. In this section, we discuss how it can actually be applied to other stages of startups.

First, because of the importance of uncertainty tolerance in seed-stage startups, the analysis in this study was limited to seed-stage startups. In other stages of startups, especially in startups that have grown more than the seed stage, the positions that require uncertainty tolerance may be different. While this ability is not required of all employees as it is in the seed stage, it is very important for managers and others in positions that require them to take into account changing market needs and trends while hiring. In addition, in startups, it is as important as in the seed phase that it is very important for the company's growth strategy to be able to easily identify managers and other qualities at the time of hiring. Therefore, this framework could be applied to startups in other stages.

Moreover, this is not limited to startups. For example, departments that create new businesses, such as new business divisions of large companies, will face a great deal of uncertainty, despite the differences from startups. In such a situation, the survey could be applied as a framework to examine whether the discussion is not overly focused on a specific topic and whether it allows for uncertainty.

6 Conclusion

This study set out to answer the question: "What abilities are essential for individuals to become integral members of an organization's core group, inspire fellow team members, and demonstrate substantial value?" Our analysis focused on identifying the distinctive traits of individuals who have remained with the organization versus those who have departed. The findings revealed several key factors that may contribute to these differences, including the ability to analyze facts, think proactively amidst uncertainty, and the frequency of contributions to methodological discussions, as highlighted through Principal Component Analysis (PCA).

Previous research has indicated that a lack of essential employee skills poses a significant risk to startup viability (Barr et al., 2009; Cummings et al., 2018; Markham et al., 2010; Thompson, 2018), especially during the seed stage, which is characterized by notable fluctuations in organizational size and associated risks (Failed, 2021; Tomczak & Brem, 2013; Ycombinator, 2022; Zhang & Wang, 2017). Additionally, divergent thinking has been identified as a critical skill for entrepreneurs, closely tied to their capacity for ambiguity tolerance and creativity under conditions of high uncertainty and risk (Lubart, 2016). Existing frameworks have provided insights into strategic focus areas, such as strategy formulation, implementation, and resource management during a startup's growth (Hu et al., 2021), emphasizing the importance of strategic leadership (Subramaniam & Shankar, 2020), strategic entrepreneurship (Morici & Zander, 2020), and adeptness in dynamic environments (Lubart, 2016; Schilke et al., 2018). However, there is a notable gap in qualitative case studies that demonstrate how these capabilities are crucial for preventing startup failures, particularly through the nonlinear and rapid application of technology. Moreover, the literature shows a lack of research on how these capabilities manifest in conversational attributes within startup organizations and the ways they can be quantified and operationalized within a framework (Lubart, 2016).

The theoretical implications of this study are significant. By focusing on a specific startup case study managed by the author, this research not only highlights the essential capabilities needed by startups (Hu et al., 2021; Lubart, 2016; Morici & Zander, 2020; Schilke et al., 2018; Subramaniam & Shankar, 2020), especially high-tech ones at the seed stage but also presents an operational framework for assessing startup uncertainties. This study emphasizes the critical role of uncertainty tolerance in startups and showcases how it translates into unique conversational traits. Our quantitative analysis, enabled by text mining, reveals that the ability to tolerate uncertainty is not merely about focusing on technical details through convergent thinking. Instead, it primarily involves divergent thinking, even among engineers, demonstrating that the ability to transcend technical jargon, hypothesize, stay mindful of basic principles without being limited by methodologies, and turn these insights into concrete actions is integral to engineers' divergent thinking abilities.

Practically, this study underscores the importance of accurately identifying such abilities in startup management and comprehending them through a framework that accommodates limited resources (Duane Ireland & Webb, 2007; Hitt et al., 2011; Hu et al., 2021). Text mining of employee conversations emerges as a valuable tool in this context. Furthermore, by visualizing the principal component vectors for two specific components derived, we can differentiate employees with varying characteristics, analyze their conversational traits, and employ this analysis as a framework. However, given that this analysis was conducted on a single company, its broader applicability to other companies with different environments and participant numbers remains a subject for discussion.

Looking ahead, there exists an opportunity for continuous analysis of these abilities within Adansons, Inc., using the developed framework. This could include monitoring how individual traits evolve over time, how they are influenced by others, and identifying convergent thinking tendencies during the hiring process to mitigate potential risks. The application of these frameworks could also aid in monitoring employee engagement trends and other relevant data, facilitating process consultation to tackle issues like unexplained resignations and improving recruitment and development practices. Additionally, exploring a simplified approach for analyzing characteristics through eigenvectors derived from the variance–covariance matrix via PCA, coupled with the dot product of this matrix and meeting data from various time points, could streamline the analysis process, enhancing its efficiency.

In addition, the sample size used in this analysis is very small and is a case study in a specific situation. While the intent of the study is to establish a methodology based on data from such a limited and specific situation, it is possible that specific biases may be included in the creation of the framework. The framework proposed here is only a possibility to analyze uncertainty tolerance in the general method (PCA), and is not dependent on the sample size of the present study. On the other hand, the very small sample size of about five participants may allow us to easily isolate the characteristic of uncertainty tolerance. In other words, in a meeting with more participants, uncertainty tolerance may appear more complex in the conversation, and caution should be exercised when applying this method in a meeting with more participants. However, PCA itself is a method that can handle an increase in the number of participants and the number of dimensions of the principal component vectors, and the flow of analysis proposed in this study itself can be considered to follow the same one. For more advanced analysis, it may be possible to increase the accuracy of PCA results by simultaneously utilizing methods to analyze components other than PCA. For example, combining with independent component analysis (ICA) may make it easier to find employees with peculiar characteristics, not limited to uncertainty tolerance, and we will consider combining such other methods with this framework in the future.

Acknowledgements We express our gratitude to the board of Adansons, Inc., for granting us access to their meeting data while ensuring the confidentiality of their information.

This study received support from JPPS KAKENHI through grant numbers JP19H01516, JP20H01546, and JP22K01742.

Appendix

See (Figs. A1, A2 and A3).

	talk	something	easy	What	Reference (patent related word)	area	me	various	...
PC1	-0.041917	0.106697	-0.038094	0.140951	0.018701	-0.026493	0.010736	-0.154798	...
PC2	0.198506	-0.185796	-0.09743	0.020601	-0.14186	-0.0787	-0.110679	0.109422	...
PC3	-0.12907	-0.057786	-0.240778	0.17922	-0.214006	-0.25301	-0.238734	0.089002	...
PC4	-0.554806	-0.082988	0.042929	-0.137654	-0.171719	0.257574	0.116944	0.095412	...

Fig. A1 Each principal component vector when PCA is applied to the July 22, 2021, meeting

	Subject4	Subject2	Subject1	Subject5	PC1	PC2	PC3	PC4
Subject4	1.000000	-0.379487	0.406635	-0.003752	-0.760231	-0.435569	0.032495	0.947770
Subject2	-0.379487	1.000000	0.234999	-0.002168	0.850822	0.118890	-0.820452	-0.142587
Subject1	0.406635	0.234999	1.000000	0.002323	-0.239442	0.530595	-0.748281	0.659330
Subject5	-0.003752	-0.002168	0.002323	1.000000	-0.001059	0.007906	-0.000015	-0.003006
PC1	-0.760231	0.850822	-0.239442	-0.001059	1.000000	0.094441	-0.444118	-0.627490
PC2	-0.435569	0.118890	0.530595	0.007906	0.094441	1.000000	-0.406975	-0.230404
PC3	0.032495	-0.820452	-0.748281	-0.000015	-0.444118	-0.406975	1.000000	-0.279105
PC4	0.947770	-0.142587	0.659330	-0.003006	-0.627490	-0.230404	-0.279105	1.000000

Fig. A2 Correlation coefficients between each principal component and each speaker when PCA is applied to the July 22, 2021, meeting

	PC1
AI	0.186293
little bit	0.183401
that's right	0.182108
voice	0.181212
research	0.180585
Oh	0.178133
Speak	0.177453
Say	0.176751
called	0.176223
think	0.175723
information	0.174867
that	0.174867
channel	0.17458
CONFIDENTIAL	0.173501
individual	0.17277
maybe	0.170156
ICA	0.170006
Dr.Kimura	0.168764
just	0.167073
can	0.163604
human	0.161232

	PC2
already	0.232055
look	0.230957
some time ago	0.229324
correctly	0.225712
yes	0.224427
professor	0.223381
person	0.2233
there is	0.220558
write	0.2189
nice	0.206794
story	0.198506
various	0.196642
exist	0.184411
now	0.168289
voice	0.14629
ah	0.137249
I know	0.129184
understand	0.128924
tolerably	0.122155
just	0.121819
try	0.118857

	PC3
oh	0.17922
human	0.132568

Fig. A3 Words comprising each principal component (threshold at 0.1) when PCA is applied to the July 22, 2021, meeting

References

Atzmon, M. B., Vanderstraeten, J., & Albers, S. (2022). Small-firm growth-enabling capabilities: A framework for young technology-based firms. *Technovation, 117*.

Barr, S., Baker, T., Markham, S., & Kingon, A. (2009). Bridging the Valley of Death-Lessons learned from 14 years of commercialization of technology education. *The Academy of Management Learning and Education, 8*(3), 370–388.

Bhattacharya, U., Hsu, P. H., Tian, X., & Xu, Y. (2017). What affects innovation more: Policy or policy uncertainty? *Journal of Financial and Quantitative Analysis, 52*(5), 1869–1901.

Blair, C. A., Gorman, C. A., Helland, K., & Delise, L. (2014). The smart leader: Examining the relationship between intelligence and leader development behavior. *Leadership and Organization Development Journal, 35*(3), 241–258.

Borracci, R. A., Ciambrone, G., & Arribalzaga, E. B. (2021). Tolerance for uncertainty, personality traits and specialty choice among medical students. *Journal of Surgical Education, 78*(6), 18851895

Byrne, K. A., Peters, C., Willis, H. C., Phan, D., Cornwall, A., & Worthy, D. A. (2020). Acute stress enhances tolerance of uncertainty during decision-making. *Cognition, 205.*

Bzhalava, L., Kaivo-oja, J., & Hassan, S. S. (2018). Data-based startup profile analysis in the European Smart Specialization Strategy: A text mining approach. *European Integration Studies, 12*, 118–128.

Clark, H. H., Fox Tree, J. E. (2002). Using uh and um in spontaneous speaking. *Cognition, 84*(1), 73–111

Corley, M., MacGregor, L. J., & Donaldson, D. I. (2007). It's the way that you, er, say it: Hesitations in speech affect language comprehension. *Cognition, 105*(3), 658–668.

Cummings, J., Reiber, C., Roeder, W. P., & Kumar, P. (2018). The price of progress–Funding and financing Alzheimer's disease drug development. *Alzheimer's Dement, 4*, Article 330343.

Dantan, J. Y., Gayton, N., Qureshi, A. J., Lemaire, M., & Etienne, A. (2013). Tolerance analysis approach based on the classification of uncertainty (Aleatory/Epistemic). *Procedia CIRP, 10*, 287–293.

Duane Ireland, R., & Webb, J. W. (2007). Strategic entrepreneurship: Creating competitive advantage through streams of innovation. *Business Horizons, 50*(1), 49–59.

Ewing, S. M., Boring, R. L., Rasmussen, M., & Ulrich, T. (2018). Text mining for procedure-level primitives in human reliability analysis. In Advances in Human Error, Reliability, Resilience, and Performance. R. Boring, M. Rasmussen, T. Ulrich, S. Ewing, & D. Mandelli (Eds.), *International Conference on Applied Human Factors and Ergonomics (AHFE), 2017*. Advances in Intelligent Systems and Computing (p. 589). Springer, Cham.

Felix, H. (2021). Company financing stages and corresponding impacts—Pinduoduo Inc. In *2021 2nd Asia-Pacific Conference on Image Processing, Electronics and Computers.*

Felsman, P., Gunawardena, S., & Seifert, C. M. (2020). Improv experience promotes divergent thinking, uncertainty tolerance, and affective well-being. *Think Write Create, 35.*

Gbadegeshin, S. A., Al Natsheh, A., Ghafel, K., Mohammed, O., Koskela, A., Rimpiläinen, A., Tikkanen, J., & Kuoppala, A. (2022). Overcoming the valley of death: a new model for high technology startups. *Sustainable Futures, 4.*

Girdauskiene, L., Venckuviene, V., & Savaneviciene, A. (2015). Crowdsourcing as a key method for start–ups overcoming Valley of Death. *Mediterranean Journal of Social Sciences, 6*(3), 795.

Greenacre, M. (2017). *Correspondence analysis in practice.* Chapman and Hall/crc.

Hitt, M. A., Ireland, R. D., Sirmon, D. G., & Trahms, C. A. (2011). Strategic entrepreneurship: creating value for individuals, organizationa, and society. *Academy of Management Perspectives*, 57–75.

Hlavac, J. (2011). Hesitation and monitoring phenomena in bilingual speech: A consequence of code-switching or a strategy to facilitate its incorporation? *Journal of Pragmatics, 43*(15), 3793–3806.

Hu, X., Zhao, R., Gao, J., Li, J., Yan, P., Yan, X., Shao, S., Su, J., & Li, X. (2021). Relationship between proactive personality and job performance of Chinese nurses: The mediating role of competency and work engagement. *Frontiers in Psychology, 12*, 533293.

Ihaka, R., & Gentleman R. (1996). R: a language for data analysis and graphics. *Journal of Computational and Graphical Statistics, 5*, 299–314. [Online]. http://www.R-project.org. [Accessed 3 December 2021].

Ilya, K. & et al. (2018). Mapping the radical innovations in food industry: a text mining study. In Higher School of Economics Research Paper, No. WP BRP 80.

Ishida, M., & Kin, M. (2012). *Corpus and text mining*" (in Japanese), "コーパスとテキストマイニング," Kyoritsu Shuppan.

Iyke, N. B. (2020). Economic policy uncertainty in times of COVID-19 pandemic. *Asian Economics Letters, 1*(2), 17665.

Knight, H. F. (1921). *Risk, uncertainty and profit*. Houghton Mifflin Company.

Lawrence, P. R., & Lorsch, J. W. (1967). *Organization and environment: Managing differentiation and integration*. Harvard University Press.

Luna-Arocas, R. & Morley, M. J. (2015). Talent mindset competency and job performance: The mediating role of job satisfaction. *European Journal of International Management, 9*(4), 421–439. https://doi.org/10.1504/EJIM.2015.066670

Lubart, T. (2016). Creativity and convergent thinking: Reflections, connections and practical considerations Bulletin of Peoples' Friendship. *University of Russia, Series Psychology and Pedagogy, 4*, 7–15.

Markham, S. K., Rei Ward, S. J., Aiman-Smith, L., & AKingon,.I. (2010). Technology entrepreneurship in the changing business environment signaling by early stage startups–US government research grants role theory in product innovation. *Journal of Product Innovation Management, 27*, 402–417.

Matta, S., Rogova, N., & Luna-Cortés, G. (2022). Investigating tolerance of uncertainty, COVID-19 concern, and compliance with recommended behavior in four countries: The moderating role of mindfulness, trust in scientists, and power distance. *Personality and Individual Differences, 186*(Part A)

McClelland, D.C. (1973). Testing for competence rather than for "intelligence." *American Psychologist, 28*(1), 1–14. https://doi.org/10.1037/h0034092

Men, L. J. (2021). The impact of startup CEO communication on employee relational and behavioral outcomes: Responsiveness, assertiveness, and authenticity. *Public Relations Review, 47*(4).

Morici, B. C., & Zander, I. (2020). Developments in strategic entrepreneurship. *Foundations and Trends® in Entrepreneurship, 16*(6), 444–513.

Parthasarathy, S. (2022). A decision framework for software startups to succeed in COVID-19 environment. *Decision Analytics Journal, 3*.

Pedregosa, et al. (2011). Scikit-learn: Machine learning in python. *JMLR, 12*, 28252830.

Radford, A., et al. (2022). *Robust speech recognition via large-scale weak supervision*. arXiv preprint arXiv:2212.04356.

Ratner, B. (2009). The correlation coefficient: Its values range between +1/−1, or do they? *Journal of Targeting, Measurement and Analysis for Marketing, 17*, 139–142.

Van Rossum, G., Drake, F. L. (2009). *Python 3 reference manual*. Scotts Valley, CA: CreateSpace.

Schein, E. H. (2016). *Organizational culture and leadership*. Jossey-Bass Inc.

Schilke, O., Hu, S., & Helfat, C. E. (2018). Quo vadis, dynamic capabilities? A content- analytic review of the current state of knowledge and recommendations for future research. *Academy of Management Annals, 12*(1), 390–439.

Stephen, H. & et al. (2022). Global entrepreneurship monitor 2021/2022 global report: Opportunity amid disruption. Expo 2020 Dubai.

Subramaniam, R., & Shankar, R. K. (2020). Three mindsets of entrepreneurial leaders. *Journal of Entrepreneurship, 29*(1), 7–37.

Sung-Soo, H., & Yang, D.-W. (2017). Analysis of research trends related to start-up using text mining. *Asia-Pacific Journal of Business Venturing and Entrepreneurship, 12*(5), 112.

Tannenbaum, A. S. (1966). *Social psychology of the work organization*. Wadsworth Publishing Company, Inc.

Thompson, J. D. (1967). *Organizations in action, The McGraw-Hill Companies, Inc.*

Thompson, S. D. A. (2018). Scientific innovation's two Valleys of Death-How blood and tissue banks can help to bridge the gap. *Journal of Business Venturing, 33*, 35–51.

Tomczak, A., & Brem, A. (2013). A conceptualized investment model of crowdfunding. *Venture Capital*, *15*(4), 335–359.

Ycombinator (2022). Stages of startups [Online]. https://www.ycombinator.com/library/Ek-stages-of-startups, [Accessed 20 November 2022].

Zhang, W., & Wang, W. (2017). Evolutionary game analysis of equity crowdfunding. In *2017 4th International Conference on Industrial Economics System and Industrial Security Engineering (IEIS)*, pp. 1–5.

Index

H. da Hora et al. (eds.), *Technology Mining*, https://doi.org/10.1007/978-3-032-10849-4

MIX
Papier aus verantwortungsvollen Quellen
Paper from responsible sources
FSC® C105338

If you have any concerns about our products,
you can contact us on
ProductSafety@springernature.com

In case Publisher is established outside the EU,
the EU authorized representative is:
Springer Nature Customer Service Center GmbH
Europaplatz 3, 69115 Heidelberg, Germany

Printed by Libri Plureos GmbH
in Hamburg, Germany